Stance devices in tourism-related research articles:
A corpus-based study

Linguistic Insights

Studies in Language and Communication

Edited by Maurizio Gotti,
University of Bergamo

Volume 292

PETER LANG
Bern · Berlin · Bruxelles · New York · Oxford

Francisco J. Álvarez-Gil

Stance devices in tourism-related research articles: A corpus-based study

PETER LANG

Bern · Berlin · Bruxelles · New York · Oxford

Bibliographic information published by die Deutsche Nationalbibliothek
Die Deutsche Nationalbibliothek lists this publication in the Deutsche National-
bibliografie; detailed bibliographic data is available on the Internet
at ‹http://dnb.d-nb.de›.

Library of Congress Cataloging-in-Publication Data
A CIP catalog record for this book has been applied for at the
Library of Congress.

ISSN 1424-8689   ISBN 978-3-0343-4555-2 (Print)
E-ISBN 978-3-0343-4551-4 (E-PDF) E-ISBN 978-3-0343-4556-9 (EPUB)
DOI 10.3726/b20000

© Peter Lang Group AG, International Academic Publishers, Bern 2022
bern@peterlang.com, www.peterlang.com

Acknowledgements

First of all, I cannot begin in any way other than by thanking Professor Francisco Alonso Almeida, coordinator of my research group, for his inestimable contributions to this project as well as for his advice, indications, and constant support.

In addition, thanks are due to the Department of Modern Philology, Translation and Interpretation and to the research team Discourse, Communication & Society of the University of Las Palmas de Gran Canaria for their financial support.

Last but not least, I would like to express my gratitude to my family and friends and especially to my partner for the immense generosity and patience shown during the completion of this project. This book would not have been possible without your encouragement.

Contents

List of abbreviations

ADJ	Adjective
ADV	Adverb
C/T	Clauses per T-unit
CN/C	Complex Nominals per Clause
CN/T	Complex Nominals per T-unit
CQL	Corpus Query Language
DC/C	Dependent Clauses
DC/T	Dependent Clauses per T-unit
DET	Determiner
ESP	English for Specific Purposes
IMRD	(model) Introduction, Method, Results, Discussion
INF	Infinitive
L2	second language
L2SCA	Syntactic Complexity Analyzer
MLC	Mean Length of Clause
NP	Noun phrase
POS	Part of Speech
RA	Research Article
SD	Standard Deviation
SFL	Systemic Functional Linguistics
V	Verb
VP/T	Verb Phrases per T-unit

1 Introduction

The study presented here is the natural evolution of my research on tourism-related research articles and other genres (Álvarez-Gil and Domínguez Morales 2018; Álvarez-Gil, Payet and Sánchez Hernández 2020; Álvarez-Gil and Domínguez Morales 2021). Despite the interest in tourism discourse in recent decades, the research article (RA) in its full form has not been the subject of extensive discursive and pragmatic research in this field. In fact, RAs on tourism as a genre have not been characterized at all (cf. Lin and Evans 2012), and this has somewhat hindered their inclusion in broader projects dealing with disciplinary discourses. However, academic research has focused on specific sections of the RA in tourism and on other genres in this domain, namely abstracts (Ahmed 2015), brochures and pamphlets (Hiippala 2015), and promotional literature (Ruffolo 2015).

The value of empirical research evidence for the tourism industry makes the study of these RAs relevant for linguistic inquiry. Much of the investigation carried out and presented as a RA concludes with a set of practical recommendations that the tourism industry and the governmental tourism offices might take into consideration to improve the situation. Lending this information in a way that sounds authoritative while protecting authors' face wants requires a great deal of rhetorical expertise. Indeed, the shift from an epistemic to effective modality to express advice and necessity implies a careful selection of language devices that are semiotically relevant for the discourse community to which these texts are addressed. In this context, an analysis of the stance-taking devices deployed in these texts seems in order. It should not be overlooked that while the primary audience of tourism RAs is scholars in the field, the presentation of practical research outcomes may suggest a wider audience, including tourist market specialists. This necessarily imposes some discourse constraints or requirements.

Surprisingly, as noted earlier, a description of the rhetorical organization of tourism RAs has not been undertaken despite the educational benefits this may bring for tourism students and non-linguistic professionals. Earlier literature (e.g., Lin and Evans 2012) comments on the apparent fuzziness of this RA structure, to the extent that it is difficult to identify major structural sequences. Based on my inspection of a corpus of RAs in the field, I argue in this volume that genre stages in these RAs present an identifiable rhetorical pattern, which may fit the traditional introduction, method, results, and discussion (IMRD) model to some extent.

As I shall explain later in this volume, the RA in the field of tourism shows some register specificities that may render methodological passages more narrative than expositive. Narration seems to play a prominent role, both to depict the research setting and to offer justificatory grounding. In this sense, language comes to evince perspectives in the elaboration and disambiguation of meaning, which, I hypothesize, represent different research momentums. These impetuses unfold linguistically differently in each RA stage to build both content and attitudinal meaning, thus reflecting the authors' stance.

A revision of the literature on stance in academic writing unveils some language devices with an indexical function. Among these studies, Chen (2009), Wang and Tu (2014), Ferdinandus (2016), and Okuyama (2020) focus on tense in different academic genres to show the way in which authors deploy this grammatical device to convey some degree of involvement in the description of the research conducted. Somehow connected to tense, the use of the passive voice in academic writing has been a matter of discussion, as the work of Ruelas Inzunza (2020) evinces. This device is frequently considered as an impersonalizing strategy to avoid accountability (cf. Rundblad 2007).

Modal verbs, especially core modals (Denison 1993), have a scope over the proposition to modulate and fine-tune the authorial perspective to meet certain communicative and pragmatic functions, as shown in Carrió Pastor (2012) and Naht et al. (2020). Modals may help mitigate, or otherwise strengthen, the propositional illocutionary force (Alonso-Almeida and Álvarez-Gil 2019, Álvarez-Gil 2021). These same functions also apply to the use of intensifiers in academic writing (cf. Pick and Furmaniak 2012).

Conditionals, as pointed out in Cacchiani (2018), serve the purpose of revealing authors' own logical thinking in the development and elaboration of meaning in a RA, as I shall show below. This argues for the evaluative dimension of these structures in the creation of new information. Likewise, *that*-complement clauses have been found to clearly show authors' evaluation of the event or phenomenon being described, as noted in Hyland and Tse (2005a, 2005b), Godnič Vičič (2015), and Kozáčiková (2021).

That said, my main objective in this volume is to evaluate the presence and function of these perspectivizing language devices in a compilation of RAs in the register of tourism. Thus, this study follows a research method based on corpus linguistics. As complex searches must be undertaken, corpus query language (CQL) is used to tailor the corpus inquiry to my needs. For this reason, the corpus is conveniently tagged for parts of speech (POS). Corpus tools will allow for the retrieval of n-gram samples as well as collocates and concordances. Statistical significance will be also assessed with the data obtained from the interrogation of the corpus. Other research objectives that need to be accomplished in order to fulfil my primary one concern the identification and description of the rhetorical organization of the RAs. This is especially relevant for detecting the statistical significance of these devices per genre stage.

This study aims, therefore, to answer the research questions below:

1. How is authorial stance exhibited in academic RAs in the domain of tourism?
2. Due to the relevance of RA rhetorical organization for this study, what macrostructure is revealed by the RAs in my compilation?
3. What are the forms and distributions of the stance-taking devices reviewed in this study?
4. Is there significant variation concerning the use of stance-taking devices among genre stages?
5. What are their overarching functions in the presentation and elaboration of meaning?

As shown by these questions, a prior step includes the identification of the macrostructure of the RAs in order to interrogate texts per genre stage. My framework of analysis considers systemic functional linguistics (SFL) (Martin 2000), especially the notion of *genre*, to provide an account of the different functional stages in which a RA may unfold and the targets these stages may have. As I will try to exemplify, there is a strong association between a stage's objectives and the language exhibited to present and negotiate meaning with potential readers.

In addition to this introduction, the structure of this volume is as follows: Chapter 2 presents my working definition of *stance*. In addition, this chapter also offers a revision of the literature on stance and describes the stance-taking devices whose forms and functions are studied in my corpus of RAs in the field of tourism. Chapter 3 presents my notion of genre and the theoretical framework with which I analyse the rhetorical organization of the RAs in my corpus. This rhetorical organization is described and exemplified accordingly. The methodology of the research and the corpus of RAs compiled for this study are included here. Along with this and to fully characterize and validate my proposed segmentation into stages of tourism research articles, I also offer some evidence based on the identification of common lexical templates per genre stage and some measures concerning informational density.

Chapter 4 presents the findings of my research and a discussion of these results. Here, information is presented quantitatively with precise statistical data concerning the evaluation of variance. A qualitative assessment of the results is also performed to unveil the discourse and pragmatic functions of stance-taking features. The last chapter includes the concluding remarks to answer the research questions given above. In order to allow replication of this study, an appendix with all the references included in my corpus of RAs is given.

2 Stance

2.1 Introduction

In this chapter, I focus on the notion of stance and on the description of the set of stance devices frequently used in RAs to indicate the perspectives of the authors, namely tenses, modal verbs, intensifiers, passives, conditionals, and *that*-complement clauses. As pointed out in the introductory chapter, these features have not been selected at random but follow from an inspection of the literature (see Gotti 2009, 2012) conducted to determine the types of devices frequently found to indicate either authorial involvement or detachment concerning the information expressed.

2.2 The notion of stance

Stance is a composite linguistic term, as it covers many language aspects and devices whose function is to designate the expression of authorial perspective. Scholarly publications on stance and stance-taking devices are numerous, even if there is still no agreement as to what the study of stance really implicates. The following definitions suggest that the way in which the concept of stance is approached is not monolithic, and it very much depends on scholarly traditions and individual interpretations. These descriptions of the concept have been taken from such sources as Biber et al. (1999), Hyland (2005), Johnstone (2009), and Dzung Pho (2013):

a)   Stance relates to the expression of speakers' and writers'
     "personal feelings, attitudes and value judgements, or assess-
     ments" (Biber et al. 1999: 966).
b)   Stance is defined by Hyland, as follows:

> [Stance] can be seen as an attitudinal dimension and includes features
> which refer to the ways writers present themselves and convey their judge-
> ments, opinions and commitments. It is the way that writers intrude to
> stamp their personal authority onto their arguments or step back and dis-
> guise their involvement (Hyland 2005: 176).

c)   "Stance is generally understood to have to do with the meth-
     ods, linguistic and other, by which interactants create and
     signal relationships with the propositions they utter and the
     people they interact with" (Johnstone 2009: 30–31).
d)   Stance is "the writer's identity as well as the writer's expres-
     sion of attitudes, feelings, or judgements" (Dzung Pho 2013: 3).

Biber et al. (1999) and Hyland (2005) stress the attitudinal side of
stance, but Hyland (2005) is more descriptive in including the notion of
authorial commitment, which does not actually seem to be at all clear
in the literature (Cornillie 2009; Alonso-Almeida 2015b). Actually,
this idea of commitment may result from the readers' interpretation
concerning a particular stance taking device, even if commitment is
not intended by the authors. (cf. Alonso-Almeida 2012). The epistemic
dimension Hyland (2005) identifies as the core stimulus for the use
of some markers, especially hedges, may support the inclusion of this
legitimizing motivation in his definition. Yet the notion of authority,
which may indeed work well with the notion of commitment, may hin-
der the epistemic integrity of markers and the way in which they may
qualify the action described in terms of the degree of likelihood of
occurrence. For this reason, Hyland (2005) also includes the possibility
that stance markers have to minimize the authors' involvement in the
propositional content given. My perspective is that commitment and
involvement do not seem to be interchangeable concepts, as they do not
point to the same linguistic, social, and psychological reality.

Johnstone's definition (2009) also reports on the attitudinal vision
of stance as an indication of the authors' and the recipients' relationship to the informational status of utterances. In the case of Dzung Pho

(2013), stance is said to involve an author's identity, but this may also suggest something about the writer's style, which does not seem to be the case. These definitions are similar in that all of them identify the evaluative dimension of stance. In general terms, stance can be understood as the way in which speakers appraise people, objects, and ideas, but it also covers self-evaluation, as Alonso-Almeida (2015a: 1) claims. Evaluation is defined by Hunston and Thomson (2000: 5) as follows:

> Evaluation is the broad cover term for the expression of the speaker or writer's attitude or stance towards, viewpoint on, or feelings about the entities or propositions that he or she is talking about. That attitude may relate to certainty or obligation or desirability or any of a number of other sets of values. When appropriate, we refer specifically to modality as a sub-category of evaluation.

This notion of evaluation leads me to consider other linguistic phenomena included within the arena of stance, namely epistemic stance (Marín Arrese 2011), epistemic modality (Cornillie 2009; Kranich 2009), commitment (Branbater and Dendale 2008), mitigation (Caffi 1999, 2007), reinforcement (Brown 2011), involvement (Cornillie and Delbecque 2008), hedging (Hyland 1998, 2005), politeness (Brown and Levinson 1987), modality and evidentiality (Chafe 1986; Diewald et al. 2009), affect (Ochs 1989), and vague language (Myers 1989; Channell 1994). All of these concepts have received scholarly attention as rhetorical devices that convey the mitigation or strengthening of claims.

Biber et al. (1999) consider the term stance to be a superordinate that covers not only the meaning speakers want to convey but also the propositional content. The term is defined as the expression of "personal feelings, attitudes and value judgements, or assessments" (Biber et al. 1999: 763), as already mentioned. Still, the notion of appraisal along with authorial attitude is highlighted in their definition, matching the other descriptions of stance mentioned in this work. As we shall see in the description of my samples from tourism articles, language variation among sections in RAs depends largely on these (inter)subjective notions leading to the selection of certain stance-taking language variables.

2.3 Stance devices

The linguistic devices that can convey stance are numerous, including tenses, modal verbs, intensifiers, passive sentences, conditional sentences, and *that*-complement clauses among others. I have selected these for inspection in tourism RAs because they have been widely used in the literature characterizing the language of RAs (Vázquez Orta 2010; Godnič Vičič 2015; Lee 2015; Ngula 2017; Sepehri et al. 2019; Díaz-Redondo 2021; Laghari et al. 2021).

2.3.1 Tense

According to Depraetere and Langford (2020: 175), tense is the device used "to establish the temporal location of a situation referred to in a clause." Brown and Miller (2016: 224) have highlighted the organizing function of tense to build up clauses and texts. The deictic nature of tense allows authors to situate events described in the text in a temporal axis so that speakers may have a referential point as to their own location in time. Thus, speakers can understand both the relationship among these events and how they logically progress over time.

In English, tense can be present, past, or future. The present tense is used to indicate events that are concurrent with the moment of speech or cognitively construed to represent a time that is logically relevant and true to the current speaker's temporal location. This explains the use of the present tense in the following instances:

(1) The present paper seeks to explore the use of modal verbs in research articles.

(2) The modal verbs in English do not inflect for person marking.

The use of the present tense in the first instance follows from the logical representation of the speaker's temporal position with respect to the time in which they elaborate meaning. In the case of the second example, the present tense simply indicates a general, accepted truth. From a semantic and pragmatic standpoint, one clear difference between the two utterances includes the notion of involvement in relation to time, as the author's participation in the elaboration of the second instance is null in contrast to the first instance, which depicts a concurrent event.

The past tense reports on events that occur before the moment of speech, as in *The companies invested in a renovation of the accommodation facilities*, where the past tense is deployed to designate an event occurring previously to the time of the utterance. Finally, the future tense shows actions that have not taken place yet and are expected to occur at a time later than the moment of speech, as in *Research will consider new demands of the tourism industry*. While the present and the past take tense morphemes, either as clitics or as vowel ablauts, the future in English is periphrastic and relies on lexical co-occurrences to show tense, such as *will, shall, to be going to + infinitive,* or the progressive, i.e., *be + V-ing*. In the case of the future with *will*, this form can also show modal overtones, as in *I will revisit the concept of dark tourism in the following pages*. In this case, volitional meaning is intended, as it is traditional in promissory speech acts like this one (cf. Brown and Levinson 1987). The action is necessarily accomplished at a time following the moment of speech.

2.3.2. Modal verbs

The concept of modality in linguistics refers to the expression of a speaker's assessment of the propositional contents regarding such notions of probability, possibility, obligation, permission, and necessity. These report on the authorial perspective concerning the information given (Palmer 1986: 2). These notions are also related to (inter)subjective uses of modal particles and to the association of modality with degrees of propositional truth, and this in turn reflects the speaker's commitment to propositional truth. The ways in which modality can be manifested in the language can be lexical or grammatical. The lexical modal devices include adverbs and other related expressions that reveal the speaker's perspective concerning the propositional content (Álvarez-Gil 2020). Palmer (1986: 33 ff) includes modal verbs, mood, and particles and clitics as examples of grammatical devices signalling modality.

Depending on the theoretical stance followed, modality can be defined and categorized in diverse ways. The most frequent division includes the categories of epistemic and deontic modality. Epistemic modality is "concerned with matters of knowledge or belief on which basis speakers express their judgements about state of affairs, events or actions" (Hoye 1997: 42). Deontic modality refers to the "necessity of acts in terms of which the speaker gives permission or lays an obligation for the performance of actions at some time in the future" (Hoye 1997: 43). The following are examples of these modalities:

(3) Luckily, Jane *may* win the prize.

(4) Jane *should* share her prize with the rest of the team.

The example with *may* in (3) reports on the chance of Jane winning the prize, hence the value of *may* as an epistemic modal verb. In the case of *should* in (4), this modal verb indicates advice to carry out the action described, and so it is an instance of deontic modality. A related taxonomy is given in Biber et al. (1999). In their classification, the terms *intrinsic* and *extrinsic modality* are deployed:

> Intrinsic modality refers to actions and events that humans (or other agents) directly control: meanings relating to permission, obligation, and volition (or intention). Extrinsic modality refers to the logical status of events or states, usually relating to assessments of likelihood: possibility, necessity, or prediction. (Biber et al. 1999: 485)

This twofold distinction covers many of the instances of modal verbs. There is, however, a third category, i.e., *dynamic modality*. This type is defined as a subcategory of Palmer's *event modality* (2001). Palmer's classification distinguishes between *propositional modality* and *event modality*. Propositional modality includes the subcategories of epistemic modality and evidentiality. While the former concerns the chances there are of a particular event to be true (Cornillie and Delbecque 2008), evidentiality specifies the evidence given for the epistemic status of the propositional content, as put forward in Willet (1988). This indexical function of evidentiality as a stance-taking device is captured in the following quote: "Stance-taking causes the speaker, among other things, to specify the sources of his knowledge, whether it be to give the utterance more weight or to reduce his own responsibility for the contents" (Haßler 2015: 183).

Event modality refers to the attitude toward events in the future. This type of modality is further divided into *deontic modality* and *dynamic modality*. Deontic modality includes senses of obligation and permission triggered especially by external rather than internal factors. Dynamic modality includes conditions that are external, and this category implies senses of willingness and ability on the part of the speaker or writer. Dynamic modality is indeed an important language feature in technical texts, as the uses of some modals can only be justified according to such notions as disposition and potentiality, as we shall see in due course. These notions have been the matter of extensive discussion in the domain of logics.

Modal meaning can be manifested by the use of certain language devices, namely modal verbs, adverbs, and clitics. In the case of modal verbs, and as pointed out in Biber et al. (1999: 483), there are nine central modal verbs in present-day English, namely *can, could, may, might, shall, should, will, would,* and *must.* Another group of modals is called *peripheral modals, marginal modals,* or, more widely, *semi-modals: need (to), ought to, dare (to),* and *used to.* This group of modals may take *to*-infinitive forms rather than bare infinitive forms. These marginal modals share some aspects with central modals, and these are (a) direct negation with *not* (also as a contracted form) and (b) inversion in questions, even if forms such as *dare* and *need* may take the periphrastic *do.* The expressions *have to, had better,* and *be supposed to* are considered as idiomatic expressions with modal overtones (Biber et al. 1999: 484).

Denison (1993: 292 ff) categorizes modal verbs from a morphological, syntactic, and semantic perspective. The criteria for the identification of modal verbs can be seen in Table 1.

Criteria
Modal verbs do not present non-finite forms.
Tense distinction takes place in the majority of these verbs.
Modal verbs do not show third-person singular present indicative suffixes.
Most modals can show their contracted version to form the negative, e.g., *can't*, *won't*, and *shan't*, and a number of these modal verbs can also appear as a clitic form, e.g., *'ll* (*will*) and *'d* (*would*).
Modal verbs do not have imperative forms.
They are followed by a bare infinitive.
Modal verbs have a scope over the propositional content.
More than one modal verb can co-occur in some dialects, and
as operators, they may share a same set of NICE properties: N = **n**egated using n't/ not, I = subject–verb **i**nversion, C = **c**ode, i.e., the modal verb is able to retrieve the meaning of an elided lexical verb in the same phrase, and E = modal verbs can be **e**mphasized.

Table 1. Criteria for the identification of modal verbs according to Denison (1993).

2.3.3 Adverbs: Intensifiers

Adverbs may be used with a perspectivizing function (Hoye 1997) in English, and some of them will be covered in greater detail in this section, *viz.,* intensifiers. There are other epistemic and evidential adverbs that are deployed either to indicate some degree of probability concerning the truth of the statement modulated, i.e., epistemic adverbs, or to suggest the status of the evidence to present a claim, as in the following examples:

  (5) *Surely* she misses her sister.

  (6) *Probably*, she misses her sister.

These two instances illustrate cases of evidential and epistemic adverbs. In the case of (5), the adverb *surely* can be used to implicate the assumption that the speaker has evidence to support a claim, hence the evidential meaning. The form *probably* in (6) reveals the speaker's lesser degree of commitment to the information presented. In this sense, this epistemic value has a hedging target in the sense given in Hyland (2005) to avoid full responsibility for the claim. As discussed elsewhere (Álvarez-Gil 2017, 2018a, 2018b), these types of adverbs have greater potential for the elaboration and negotiation of scientific knowledge, as I will show at the proper time in this volume.

In general, adverbs are a rather heterogeneous grammatical category, and this characteristic adds to the complexity of providing a straightforward description of the adverb category. Furthermore, this reality has been referred to by various scholars, such as van der Auwera and Plungian (1998), Haspelmath (2001), and Eisenberg (2013), who have stated that the grammatical category of adverbs is the "most problematic class of head words" (Haspelmath 2001: 16543) and that it is an "elusive" (van der Auwera and Plungian 1998: 3) and sometimes "confusing" (Eisenberg 2013: 212) part of speech. The absence of a clear definition, therefore, means that for some scholars, this category serves as a catch-all for all lexical items that do not qualify for inclusion in other categories.

One of the reasons for this is the diversity of adverbs from a formal perspective, as some words look like adverbs. This is the case with the adjective "daily," for example, since it shares the *-ly* ending with adverbs. In addition, an adverb can take up various positions in a clause depending on the intended meaning and the syntactic function the adverb fulfils, which complicates the assignment of grammatical category, as Huddleston and Pullum (2002) point out. The examples below serve to illustrate these points.

This ending *–ly* traces back to the Old English suffix *–lic* in adjectives, as "historically adverbs in English were inflected forms of adjectives" (Fontaine 2013: 33). Hence, today, it is generally related to the grammatical category of adverbs. There are still some cases of *–ly* adjectives, as in the following examples:

(7) The night seems *chilly*.

(8) The river water became *bubbly* after the earthquake.

Some adjectives and adverbs may show the same form, e.g., *daily* and *fast*.

(9) *Daily* visitors are welcome at the hotel entrance (adjective).

(10) We welcome visitors *daily* at the hotel entrance (adverb).

(11) I will get an *early* flight to Madrid (adjective).

(12) The train arrived *early* (adverb).

(13) He is a *fast* runner (adjective).

(14) He runs *fast* (adverb).

The above instances exemplify how such words can function as adjectives and adverbs depending on the syntactic role they fulfil, even if these forms are semantically related. There are also cases in which the semantic meaning of an adjective is different from that of its corresponding adverb form. That happens with the word *dead*, whose meaning as an adjective is *to be no longer alive,* while the adverb is used as an intensifier to enhance the meaning of the adjective, as in *This is dead good.* The adverb *dead* turns to signify *perfectly/totally.*

From a morphological standpoint, the relationship between adverbs and adjectives is clear, as many adverbs are formed through derivation by the addition of certain suffixes. Some common suffixes besides –*ly* in English are the following: –*ward*(s), as in *afterwards* (preposition + –wards* [suffix]); and –*wise*, as in *clockwise* (noun + –*wise*) and *nowise* (determiner + –*wise*). These last instances show how the suffix –*wise* can modify the meaning of a word category to turn it into an adverb.

These nuances have made the definition of the adverb as a grammatical category a matter of ample discussion. Not in vain, Haspelmath (2001: 16543) has described the adverb as the "most problematic major word class." Traditional definitions are insufficient, as they are based on structural concerns alone. Despite these scholarly considerations, there are some successful characterizations of the term. One of this is given in Ramat and Ricca (1998: 187) and encompasses both formal and functional aspects:

i. Formally, adverbs are invariable and syntactically dispensable lexemes (which may have derivational status, e.g., Lat. *Simil-i-s* 'similar' *simil-iter* 'similarly', or even originate from inflectional status: Lat. *Merito(d)* 'rightly').

ii. Functionally, adverbs are modifiers of predicates, other modifiers or higher syntactic units. In other words they add information to other linguistic elements which can stand on their own, semantically as well as syntactically.

In this context, a further distinction needs to be made. This involves those adverbs that are integrated in the clauses and function as modifiers and the cases in which they occur on their own as independent elements in these clauses. The latter are known as adverbials, whose main characteristics as described in Biber et al. (1999) are included in Table 2.

Adverbials
They can generally be added more or less independently of the type of verb.
They are generally optional in the clause structure.
They are characteristically realized by adverb phrases, prepositional phrases, or clauses.
They are more mobile than the other clause elements, often occupying a variety of positions in the clause.
Their positions are determined to a larger extent by textual and pragmatic factors than by the positions of other clause elements, which are more determined by syntax.
Unlike the other clause elements, more than one adverbial may co-occur in the same clause.

Table 2. Main characteristics of adverbials according to Biber et al. (1999: 131–132).

The following examples illustrate the use of adverbs as modifiers. In (15), the adverb *certainly* modifies the adjective *achievable*. In (16), the adverb *consistently* modifies the verb *demanded,* and finally, in (17), the adverb is modifying another adverb, i.e., *very* modifies *seldom*.

(15) The recovery of tourism is *certainly achievable.*

(16) Local politicians have *consistently demanded* new governmental migration policies.

(17) Trains *very seldom* arrive late.

In contrast, in the following occurrences, (18) and (19), the adverbs are independent elements in the clauses they are inserted into, and they function on their own. They modify the sense of the whole utterance, while in the previous cases, the adverbs affected only the verbs they accompanied:

(18) *Luckily,* the applicants' response was positive.

(19) *According to* the latest research, some measures should be taken to control gas emissions from laundries in tourist areas.

As noted in Biber and Finegan (1988), adverbials have a patent interpersonal function. Adverbials can be classified into *adjuncts, conjuncts,* and *disjuncts* (Greenbaum 1969; Quirk et al. 1972; 1985) or into *circumstance, stance,* and *linking* adverbials in the case of Biber et al. (1999: 763). Halliday et al. (2004: 123 ff), with a systemic functional framework, categorize adverbs into *circumstantial* or *adjunct, conjunctive, conjunct, linking,* and, finally, *modal* or *disjunct* adverbials. *Adjunct* or *circumstantial* adverbials contribute to referential meaning. *Conjunct* or *conjunctive/linking* adverbials have connective and text-organizing targets. Finally, *disjunct* or *modal* adverbials are deployed to indicate appraisal of the propositional information.

As in Biber et al. (1999), Hyland's (2005) stance adverbs clearly indicate an author's perspective toward their texts, and their use depends on the effect an author is seeking to have on readers. They can serve to express possibility or a lack of complete commitment to the truth of a specific proposition, thus exhibiting a hedging function. That is the case of intensifiers. An intensifier can be a noun, an adjective, or an adverb and "scales a quality, whether up or down or somewhere between the two" (Bolinger 1972: 17). Intensifiers are classified as *boosters, compromisers, diminishers,* and *minimizers* to refer to the upper part, middle part, lower part, and lower end of the scale, respectively, according to Bolinger (1972: 17), from which I excerpt the following instances:

> BOOSTERS: He is a *perfect* idiot. She is *terribly* selfish.
>
> COMPROMISERS: He is *rather* an idiot. She is *fairly* happy.
>
> DIMINISHERS: It was an *indifferent* success. They were *little* disposed to argue.
>
> MINIMIZER: He's a *bit* of an idiot. I don't care *an iota* for that.

In this work, I have opted for keeping two categories only in my analysis of intensifying adverbs: boosters and the rest, which I refer to as downtoners in the sense of Hyland (1998: 135) (i.e., "[s]emantically, epistemic adverbs function principally as adjuncts and disjuncts"). Downtoners include, therefore, the subcategories of compromisers, diminishers, and minimizers.

Another aspect concerning the description of adverbials is placement. Adverbs can be given in the initial, pre-verbal, post-verbal, and final positions. The places that adverbs occupy in the sentence may affect the sense of the information presented. The initial position is the first item in a clause, as *unexpectedly* in the following instance:

(20) *Unexpectedly,* she turned up.

The pre-verbal position is between the subject of the clause and the main verb, as in the case of *often* in (21) below, or between the auxiliary or the modal and the lexical verb, as in (22). The final position is exemplified in (23).

(21) Tourists *often* apply for a VAT refund.

(22) Tourists may *actually* apply for a VAT refund.

(23) Tourists can apply for a VAT refund *immediately*.

Viewpoint or stance adverbs are usually placed at the beginning, and they have an effect on the whole sentence, as in the following instance:

(24) *Evidently,* tourists can apply for a VAT refund.

Aijmer (2016) explains that a change in the position of an adverb may lead to variations in its pragmatic functions. She uses the form *actually* to illustrate that this adverb has different effects on the meaning depending on its position in the clause. This semantic flexibility of adverbials has also been described in Beeching and Detges (2014: 1); they also consider the adverbials' right or left periphery location to describe their functions.

2.3.4 Passives

The passive voice is believed to be extremely common in scientific writing, as pointed out in van Gelderen (2010: 8). For this scholar, the use of the passive in contrast to the active is a matter of style, as the passive voice has a clear communicative and pragmatic function: "They are often advised against for reasons of style because the author may be seen as avoiding taking responsibility for his or her views" (van Gelderen 2010: 8). Indeed, Biber et al. (1999: 476) have found evidence of this tendency in academic texts in order to give "topic status to the affected patient." This is possible for the transitive nature of the verbs in the active–passive constructions exchanged, as in the following examples:

(25) The manager of the damaged hotel accommodated tourists in other hotels during the storm.

(26) Tourists from the damaged hotel were accommodated in other hotels during the storm.

As to the formal aspect of the passive, Leong (2014: 6–7) identifies the below patterns (my examples).

Type	Pattern
Basic	*be* + V*en* (see example 27 below)
Progressive	*be* + *being* + V*en* (see example 28)
Perfective	*have* + *been* + V*en* (see example 29)
Modal	modal + *be* + V*en* (see example 30)
Modal perfective	modal + *have* + *been* + V*en* (see example 31)
To-infinitive	*to* + *be* + V*en* (see example 32)
Non-finite *–ing*	*being* + V*en* (see example 33)
Bare	V*en* (see example 34)

Table 3. Passive patterns.

(27) Tourists *were informed* about the hotel facilities.

(28) Tourists *are being informed* about the hotel facilities.

(29) Tourists *have been informed* about the hotel facilities.

(30) Tourists *should be informed* about the hotel facilities.

(31) Tourists *should have been informed* about the hotel facilities.

(32) The number of tourists *to be accommodated* during the storm was high.

(33) The number of tourists *being accommodated* during the storm was high.

(34) The manager met the tourists *accommodated* during the hurricane.

The deictic nature of the passive voice has been highlighted in Ding (2002). This scholar claims that its use seems to promote ideas and information centred on the work, or the object of work, rather on a person. By giving the work or the object of work a thematic position, other scholars are able to replicate and verify the findings (Ding 2002: 152). As the passive voice is a common feature in scientific writing, it is assumed that this device is not a matter of stylistic choice but a widespread and conventional way of providing expected information in scientific writing within a particular community of practice (cf. Leong 2021).

2.3.5 Conditionals

Conditional structures in English are also attested forms to express deictic content, as they have a modulating effect over the proposition in the sense that epistemic qualification is conveyed, as pointed out in Nuyts (2001: 352). The conditional sentence is formally identified as a two-part structure comprising *protasis* and *apodosis*. The protasis is the clausal element introduced by *if*, while the apodosis is the matrix clause, as in the instance below:

(35) If the storm damages the hotel, the manager will accommodate tourists in other hotels.

In (35), the protasis is the material following *if*, and the apodosis is the clause given after the comma. For Huddleston and Pullum (2017: 738), the complete sentence is a conditional construction, while the protasis with *if* represents a conditional adjunct. This interpretation assumes *if* to be the head of the prepositional phrase, contrary to traditional grammars, in which *if* is a subordinating conjunction. Ignoring my own caveats concerning the proposal in Huddleston and Pullum (2017), the example in (35) is understood as an instance of a conditional sentence where *if* is part of the protasis functioning as a subordinating particle, as is the view in many modern grammars. Dapraetere and Langford (2020: 297) argue that the subordinator *if* also appears "in embedded questions as an alternative for *whether,*" as in (36):

(36) The manager was asked if tourists would be accommodated in other hotels during the storm.

Biber et al. (2021: 811–820) classify conditions in a clause from a semantic point of view into open conditions, hypothetical conditions, and rhetorical conditions. Biber et al. (2021: 812) provide the following instances to illustrate these categories:

(37) The annual tourism convention will not be able to take place this year if air traffic controllers eventually go on strike.

(38) If a decree could have been enacted to limit the new construction of hotels, this would have prevented land congestion on the part of the large hotel companies.

(39) You may think that I want to destroy the milk boards, but if you believe that you will believe anything.

The first instance reports on a real condition whose fulfilment is unknown, and it belongs to the open condition category. The example in (38) describes an unreal situation and falls into the category of hypothetical conditions. Finally, the example in (39) is a case of rhetorical condition, meaning that clauses such as these "take the form of a conditional, but combined with the main clause, they make a strong assertion" (Biber *et al.* 2021: 812). Sentences like these ones show the potential of conditional sentences to signal stance.

Sweetser (1990: 113) offers a functional classification of conditional sentences: content conditionals, epistemic conditionals, and speech act conditionals. Content conditionals refer to a condition that may be accomplished given the fulfilment of the condition expressed in the protasis. One example could be *She will accept the invitation if her husband apologizes first*. This means that the acceptance of the invitation happens as long as the husband's apologies take place. Epistemic conditionals refer to logical correlations between the phenomena described in the protasis and in the apodosis based on one's knowledge of their possible realizations. The following instance illustrates this:

(40) If local governments were to increase the tourist tax, there would be a decline in visitors from some EU countries.

In this example, the truth of the apodosis depends on the knowledge one has about the truth of the protasis. Sweetser (1990: 116) specifically includes tautological conditionals in this group and gives the following example:

(41) If she's divorced, (then) she's been married.

This instance illustrates tautological conditionality based on actual knowledge of the world. Thus, the fact that the woman is divorced entails the fact that she must have been married first. Finally, speech act conditionals are described as follows: "The performance of the speech act represented in the apodosis is conditional on the fulfillment of the state described in the protasis (the state in the protasis *enables* or *causes* the following speech act)" (Sweetser 1990: 118).

This is exemplified in (42) and (43) below:

(42) If I may, I'd like to help you.

(43) If you don't mind, I need to go in first.

In these examples, the completion of the actions described in the apodoses depends on the realization of the speech acts given in the protases. The pragmatic function of this type of conditional falls into the domain of politeness and the avoidance of imposition. As we shall see in due course, conditionals are involved in the process of logical argumentation and in the elaboration of meaning in RAs.

2.3.6 *That*-complement clauses

The evaluative dimension of *that*-complement clauses has been reported in Hyland and Tse (2005a; 2005b), Hyland and Jiang (2018), Kim and Crosthwaite (2019), and Alonso-Almeida and Álvarez-Gil (2021b) among other works. As pointed out in Charles (2007), these structures have a strong perspectivizing potential, as the matrices introducing these clauses may convey meanings referring to such notions as authority, probability, accountability, attribution, affectivity, and advisability to mention a few. Hyland and Tse (2005a) highlight the interpersonal scope of *that*-structures to negotiate meaning as authors develop their ideas throughout their papers. Even so, these authors have found that these structures have not received the attention they deserve

(Hyland and Tse 2005a: 40). These devices are not only useful to indicate perspective, which is what matters in this study, but they are also important in the overall organization of the contents to provide a logical unfolding of arguments (Alonso-Almeida and Álvarez-Gil 2021a, 2021b: 229, 2021c).

Formally, *that*-complement clauses are hosted by a matrix that introduces them. These subordinating matrices present different realizations from a structural point of view, using copula verbs (*It is* [. . .]) or lexical verbs (*I think* [. . .], *They noted* [. . .], *We witnessed* [. . .], *The author has reported* [. . .]). Semantically, the matrix both indicates the evaluation and is the source of evaluation, as described in Hyland and Tse (2005a: 40) and Hyland and Jiang (2018: 140), while the *that*-clause is the evaluated entity, as shown in the following example taken from the latter source:

(44) [evaluation & source of evaluation ➜] I will show [evaluated entity ➜] that the biofunctional account can give us more content specifity than Fodor supposes.

The evaluation of the entity can be realized by means of a noun or an adjective, as is the case with *evidence* and *possible* denoting differing degrees of epistemic justification in the following instances:

(45) There is *evidence* that traders may reduce their cost from trading by splitting orders so as to dampen the pressure on inventory holding (Hyland and Tse 2005b: 125).

(46) It is *possible* that mainstream teachers overcompensate and are especially lenient with NNSS (Hyland and Tse 2005a: 42).

Hyland and Tse (2005b: 46 ff) note that these structures can be categorized according to the entity evaluated, the author's stance, the source of evaluation, and the formal aspect the evaluation is presented with. These categories, in turn, report on semantic subcategories to specify the intention of the author. The entity evaluated can include "an interpretation of the writer's claim; of the content of previous studies; of research goals; and of the research methods, models, or theories that had been drawn on" (Hyland and Tse 2005a: 46). In this group, another subcategory involves "common or accepted knowledge," as explained in Hyland and Jiang (2018: 146). The function of these categories concerns the evaluation of the propositional content introduced by the matrix, e.g., *The general belief in this respect is that measures should be issued to control illegal entry of exotic animals in Europe.* In this instance, the evaluation reports on a theory or proposal, which is supported by means of the intersubjective device *The general belief is* (. . .) introducing the idea of accepted knowledge.

The type of stance devices authors use may be (a) attitudinal, (b) epistemic, and (c) neutral concerning the propositional content introduced by the lexical matrix. The neutral category includes such instances as *It means that* (. . .) (Hyland and Jiang 2018: 145–146). The attitudinal stance devices relate to aspects of affect and obligation, while the epistemic ones appraise differing degrees of likelihood concerning the propositional content. This means that for Hyland and Tse (2005a: 46), the epistemic category informs on notions concerning the truth and the accuracy of the information presented.

According to Hyland and Jiang (2018), the source of evaluation in the matrix may be humans, abstract entities, or unidentified. In the case of humans, attribution can be given to the author (*I*) or a third party (*they, the researchers, scholars*). Alonso-Almeida and Álvarez-Gil (2021b: 230) suggest "a third category, i.e., the author and colleagues (*we*) so that we can specify those cases in which the subject or subjects of conception may report on subjective or intersubjective meanings." The source of evaluation can imply such abstract entities as *the study, the report,* or *the survey,* for instance. Attribution can also be disguised by the use of opaque conceptualizers (e.g., *It is commonly accepted* [. . .]). In this case, Hyland and Tse (2005a: 46) think that authors may be purposely evading accountability for the information given. Finally, the evaluative expression refers to the forms in which evaluation is conveyed in the matrix. These could be non-verbal predicates (noun and adjectival predicates) and verbal predicates. The latter concern research acts (e.g., *The results indicate that* [. . .]), discourse acts (e.g., *The researcher states that* [. . .]), and cognitive acts (e.g., *The authors believe that* [. . .]). This resembles Marín-Arrese's (2009a, 2009b, 2011) model for the analysis of epistemic stance-taking devices.

2.4 Summary

In this chapter, I have defined the concept of stance, as it will be applied to the identification and analysis of perspectivization devices in the corpus of tourism-related RAs. This attitudinal notion is revealed as a complex one, as it includes a wide array of forms that need to be disambiguated according to the co-textual and contextual environment in which they appear (Nadova 2015). The study of stance in RAs in the field of tourism additionally requires the consideration of the genre and register variables because these may explain the presence of certain stance-taking forms in my corpus, as I shall show in the remaining chapters.

I have also described a set of linguistic features commonly associated with the expression of the author's point of view in scientific writing. These features have other ancillary effects on the elaboration of meaning, as their use involves a number of pragmatic functions that are contextually defined. The evaluation of these perspectivizing forms in my corpus of texts in tourism should show the way in which they may contribute to building on new knowledge as well as to creating textual ties.

3 The research article in tourism-related studies

3.1 Introduction

In this chapter, I focus on the rhetorical organization of RAs in the field of tourism in terms of their macrostructure. The RA has been defined as "a codification of disciplinary knowledge" (Hyland 2004: 64) and "is often considered to be the key genre of modern knowledge-creation" (Hyland and Tse 2004: 42). While the study of the RA as a broad genre has received widespread research attention, this is not the case for RAs in the field of tourism; in fact, I have not been able to find studies on this specific subject. This may be due to the apparent lack of standardization of the sections included in tourism RAs, as noted by Lin and Evans (2012). These authors further state that in the discipline of tourism, along with other disciplines such as applied linguistics and systems engineering, "no or few 'major' structural patterns could be identified" (Lin and Evans 2012: 157). This chapter therefore seeks, albeit tentatively, to remedy this situation by providing an account of the generic structure of a selection of papers published in leading journals on tourism. Alongside this, I also examine the language used in these papers—first, to justify the proposed generic segmentation and, second, to measure the language variation among the stages of the genre.

My notion of genre is based on functional linguistics, particularly the works of Martin (1984), Biber (1988), Swales (1990), and Eggins (1994) among others. Although the notion of *move* (Swales 1990) is generally accepted in the description of textual genres, I here use the concept of *stage* as described by Martin (1984: 25), although both terms refer to almost the same structural unit in genres. For my analysis of the language variables in these texts, I rely on the works of Halliday and

Martin (1993), Biber (1988), Hyland and Jiang (2018), Hyland and Tse (2005a), Holtz (2011), and Biber and Gray (2011) among others. I seek to determine how the use of all of these language variables methodically contributes to developing argumentation and signalling authors' perspectives in an attempt to negotiate meaning with their readers. In so doing, I also wish to demonstrate that despite current scholarly caveats reporting a lack of uniformity among tourism research papers, there is indeed an underlying identifiable textual structure that is also signalled by reiterating linguistic cues.

3.2 Genres and the research article

The study of RAs from a genre perspective has attracted significant attention throughout the last four decades. This is especially true for certain registers, such as medicine and engineering, but not for others, e.g., tourism and psychology (cf. Lin and Evans 2012). In particular, research into the rhetorical organization and language of complete RAs is notably less comprehensive, although there are some exceptions (Kanoksilapatham 2005; Li and Ge 2009). Moreover, the language and organization of individual RA sections appear to be well documented (Brett 1994; Gil-Salom and Soler-Monreal 2009; Loi 2010; Benkenkotter *et al.* 2012; Kawase 2015; Liu and Buckingham 2018).

Traditionally, the study of a genre has been approached from three mainstreams of thought (cf. Hyon 1996), namely new rhetoric (Bazerman 1988, 1994), English for specific purposes (ESP) (Swales 1990), and SFL (Halliday 1978; Halliday and Martin 1993). Even if every school has its own clear foundations and motivations concerning the notion of genre, in recent years, the boundaries among these schools have become less distinct and neat, with complications arising from "integrative and cross-over work," as pointed out in Tardy and Swales (2014: 166). While SFL provides a rich description of textual and language features establishing social semiotic worlds, ESP genre research offers insights into the functional and social foregrounding of genres, and new rhetoric genre research provides generous ethnographic information on the underlying construction of genres. A common aspect of these schools is their educational and pedagogical interest in seeing how genres enact particular professional and institutional practices (Gotti 2017).

The textual representations of genres constitute basic social semiotic structures in interactions, which seek to entertain certain verbal or material responses within a particular discourse community in specific contexts (see Gotti 2003, 2021). As put forward in Lin (2020: 12), drawing on Paltridge (2001), genres thus have to be identifiable by the members of a community for the correct interpretation of the communicative intent. In this sense, Lin (2020: 12) adds that these members "place constraints on what can be accepted generally in terms of textual content, register features and forms for a particular genre," each of these representing one semiotic layer of knowledge construal, including the interpersonal layer covered in this paper. The notions of purpose, content, and form are core to the definition of genres. Genres enact particular situated communicative events, and variation may emerge out of specificity concerning actual social needs. Thus, the exchange of academic research knowledge may take the form of a RA, but the textual realization of the genre enacting this social process may differ synchronically according to specific situated communicative demands. Diachronic variation is, for the same reason, possible.

The ESP and SFL research traditions see genres as structured constructs representing the unfolding of social and cultural practices. From a systemic functional perspective, Martin (1984: 25), and later Martin and Rose (2008: 6), describes a genre as "a staged, goal-oriented, purposeful activity in which speakers engage as members of our culture." The association of these stages gives way to the generic structure potential, i.e., the representation of all possible stages in a genre, and the actual generic potential, i.e., the representation of the stages occurring in a particular sample, as described in Hasan (1995) and Eggins (1994). The Systemic Functional Linguistics stages seem to correspond structurally to article sections (cf. Swales 1990); this concept is also deployed in Nwogu (1997: 125).

Based on the notion of generic structure to express experiences, Bhatia (1993: 53) sees "the concept of genre as a dynamic social process." This claim could not have been applied to RAs in a timelier fashion, particularly as regards those disciplines that have existed for a relatively short time, such as tourism. The field has a lifespan of almost two decades, as put forward in Taillon (2014: 1): "Tourism as a field of study is a new addition to academia. Until the 1990s tourism was not an accepted field of research as a standalone academic community."

The RA as a genre has been defined as "the culmination of an involved process of research" (Tessuto 2015: 13), and its function is to "report research findings derived from direct observation or various kinds of experimental studies," as noted by Lin (2020: 20), echoing Weissberg and Buker (1990). The traditional IMRD model (Swales 1990) has worked well in the rhetorical description of what Swales (2004: 13) classifies as the empirical RA. This rhetorical organization of the RA has also been a focus of study for Brett (1994), Nwogu (1997), Posteguillo (1999), Piqué and Andreu-Besó (2000), Kanoksilapatham (2005), Li and Ge (2009), Lin and Evans (2012), Basturkmen (2012), Stoller and Robinson (2013), and Tessuto (2015), *inter alia*.

In the case of the RA in the field of tourism, as noted in the intro-duction section, I have been unable to find a single study dealing with its internal organization in terms of moves, sections, or stages. Ahmed (2015: 274) has, however, studied tourism RA abstracts and identified nine sections. However, I did not find such a richness of stages in my compilation of RAs. My description of the tourism-related RA mac-rostructure in the methodological section is based on Martin's (2000) notion of staging, even if tourism texts seemingly lack a standard-ized arrangement of content, as noted in Lin and Evans (2012). In the absence of previous studies on the structure of tourism-related RAs, I will exclusively divide the RAs into stages. This division will be sup-ported by language formulae, with statistically verified associations with each genre stage.

The realization of genres, according to the SFL tradition, relies on a set of choices that evince the relationship between language and social context as semiotic systems in which the construal of meaning develops (Martin 2000: 5). An essential part of this meaning construal is the way in which individuals engage with texts and "are positioned and repositioned socially" (Martin 2000: 10). The personal imprint is thus unavoidable in the manifestation of genres, as each individual has their own ideological traits that engender a subjective position no matter how institutionalized a particular social practice might be. In the stratified account of language SFL represents, subjectivity may rely primarily on certain variables pertaining to lexicogrammar, which, in turn, redound in certain discourse semantic aspects.

It is apparent that language evincing social context reflects on certain linguistic resources. In a diversified model of register, these resources refer to representation, interaction, and information flow (Martin 2000: 4), mirroring the ideational, the interpersonal, and the textual metafunction, respectively. In this framing, subjective meaning seems to be unambiguously enacted in the expression of interpersonal linguistic cues in the unfolding of texts. Building on previous works on the language of science (Warchał 2010; Abdollahzadeh 2011; Hu and Cao 2011; Alonso-Almeida 2015b; Jaime and Pérez-Guillot 2015; Conrad 2018; Hyland and Jiang 2018; Bongelli et al. 2019; Kim and Crosthwaite 2019), I have selected certain perspectivizing language devices, namely tenses, modal verbs, intensifiers, passives, conditionals, and *that*-complement clauses, to further discuss their use and variation per genre stage in tourism-related RAs. These features pursue the implementation of an interactive dimension that is central to the elaboration of meaning within scholarship circles.

3.3 The form and linguistic characterization of the RA in the field of tourism

In this section, prior to the detailed analysis and discussion of the stance language features of tourism RAs in the next chapter, it seems appropriate to provide an account of the textual genre the RAs in my corpus represent. As discussed in the previous section, even though tourism RAs have been the focus of some research, there has surprisingly been no examination and subsequent description of the genre. Scholarly interest has focused primarily on the rhetorical organization of RA abstracts and brochures in tourism (Mongkholjuck 2008; Yui Ling Ip 2008; Ahmed 2015; Álvarez-Gil and Domínguez-Morales 2018). In the following subsection, I pay attention to the generic structure of tourism

RAs in terms of their functional stages. The information provided corresponds to my own inspection of the corpus and the authors' own labelling of stages. In addition, I support my stage segmentation of the papers under focus with a report presenting the language templates most frequently associated with each of the stages in a tourism RA. Finally, I have also measured lexical and syntactic complexity across all genre stages. These analyses may aid in the understanding of the differences in variation among the RA stages in terms of their lexical density and syntactic elaboration.

3.3.1 Methodology and corpus

The data for the present study was drawn from 74 tourism RAs published in journals between 2015 and 2018. These journals were selected in line with criteria based on the notions of indexing and visibility in international databases, as these are typically reliable means of learning about the quality of the publication process, specifically (a) the quality of the contents and article layout, (b) the established policy of malpractice and ethics, and (c) the edition. For the indexing, the criterion was the journal's inclusion in the Social Sciences Citation Index; for the visibility indicator, moreover, the criterion was a score of at least 10.5 according to the ICDS rating (*índice compuesto de difusión secundaria*, i.e., secondary composite index broadcasting) in the MIAR database (http://miar.ub.edu). This score suggests that the journal is included in several databases, one of which is within the *Web of Science Core Collections*©. Based on these parameters, I selected my texts from the following three journals: *Journal of Vacation Marketing, Tourism Economics,* and *Tourist Studies.* The examples from (47) onwards all belong to the compiled corpus of articles published in the journals mentioned.

The total number of words in these articles comes to 547,623, excluding the RA abstracts and bibliographic references. The texts were extracted from PDF files; these were in turn converted into plain text using UTF-8 encoding and stored as .txt files for use with corpus tools. This was achieved using CasualTextractor 1.0.5 by Yasu Imao (2020b) to ensure the safe extraction of the text without misspellings and incorrect word splits.

Finally, in order to retrieve data for each genre stage, each RA was carefully read to identify the stages in question. After this identification process was complete, the text corresponding to each stage was saved in separate files and stored in folders along with other files containing the same type of genre stage. Interrogating the corpus thus divided, I was able to support my stage segmentation of the papers therein with a report presenting the language templates most frequently associated with each of the stages in a tourism RA. In addition to frequent language templates, I also measured lexical and syntactic complexity across all genre stages as a justification of my segmentation of the RAs.

I also used several different software applications for text analysis and retrieval. In the case of the values concerning lexical density and syntactic complexity, these were obtained using Lexical Complexity Parser (LCA) (Ai and Lu 2010; Lu 2012) and Syntactic Complexity Analyzer (L2SCA) (Lu 2010, 2011; Ai and Lu 2013; Lu and Ai 2015), respectively. My intention was to assess variation among stages in the RAs, examined in terms of syntactic elaboration and sophistication. To this end, I chose seven out of 14 possible indices included in the automatic analysis provided by the L2SCA involving aspects of production unit length, amount of subordination, and particular structures (cf. Lu, 2014: 137–138), as we shall see below.

Another software tool used was CasualConc, developed by Yasu Imao (2020a), to obtain concordances of occurrences in unformatted text in each genre so as to excerpt examples from the corpus. I also used the Lancsbox suite (Brezina et al. 2015; Brezina et al. 2020), which contains several tools for use in interrogating a corpus (keywords, n-grams, graphical collocations, frequency lists, etc.) and calculates statistics of findings (Brezina 2018). Lancsbox also allows for part-of-speech searches, which are made possible through the prior tagging of all tokens in a corpus using the Penn Treebank part-of-speech Tagset, tokenization, and lemmatization. Wildcard, smart, and CQL searches can also be carried out with this application. All of these potentialities make it possible to conduct complex searches in order to obtain cases of (for example) extended noun phrases, premodification, and passives.

3.3.2 Genre stages in tourism-related RAs

My inspection of the articles in terms of their internal segmentation reveals the distribution of stages presented in Figure 1. All papers in my corpus present an abstract; however, this genre falls beyond the scope of the present study, as abstracts have conceptual and structural independence (i.e., they can stand alone). The abstract in tourist RAs has been fully studied in Ahmed (2015) and Sabila and Kurniawan (2020) and has been found to constitute a genre in itself rather than a stage.

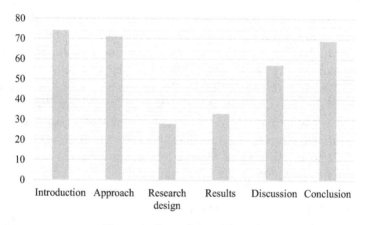

Figure 1. Stages in tourism RAs.

In the case of the RA introduction, this stage appears in all analysed RAs, even if the section was not labelled at all. The next most frequently occurring sections are the approach and the conclusion sections, followed by the discussion and the result design sections, in that order. The least frequently occurring section is the research design section. The functions of the sections are defined in Table 4.

Stages	Function
Introduction	To state the reason for writing the paper and the topic of the research.
	To state the objectives of the paper.
	To state the hypothesis.
	To put forward the theoretical tenets and methods used in the development of the research.
Approach	To describe the theoretical background for the research being conducted.
	To describe similar research activity in the field.
Research design	To describe how the research activity has been carried out; this includes the materials, subjects, and places involved, along with the procedure(s) used.
Results	To summarize the main findings from the analysis performed on a body of data (for example, phenomena observed in a particular tourist destination).
Discussion	To interpret the possible meanings of the data obtained from the analysis of results in the context of the theoretical framework.
	To provide inferences in light of the results.
	To offer concluding remarks, especially in cases where a proper conclusion stage is not included.
Conclusion	To outline the main contribution(s) of the paper and the significance of the research.
	To show authorial perspective with respect to observed data and within existing knowledge.
	To identify the strengths and shortcomings of the research presented.
	To outline directions/statements for future research.

Table 4. Functions of stages.

In addition to these stages, one of the articles in my corpus presents a section after the conclusion entitled "postlude," and this means that not much information can be gleaned from it. Its function is to offer unexpected relevant information obtained after the paper and the research under discussion had been completed. This stage seems to tentatively indicate the implications of the study for the matter under study. This is irrelevant in statistical terms and constitutes an outlier in the corpus.

In general terms, the manual analysis of the macrostructure of the 74 papers evaluated was not straightforward for one key reason. Specifically, the labelling of the RA stages in the corpus did not seem to follow a unified set of criteria; in fact, rather than conveying functional information to indicate what the stage was for, some section headings in some of the examined cases may have been descriptive of the content to the extent that headings became very creative and original. Variations in this respect are presented in Table 5.

Here, the categories *functional* and *narrative* may be roughly compared to Ruiying and Allison's (2003: 369) "conventional functional headings," "varied functional headings," and "content headings." My use of the term *narrative* rather than *content* headings reflects the rhetorical elaboration exhibited by these headings. Furthermore, functional headings, no matter how traditional, are also content headings, as they clearly suggest meaning.

Sections	Headings		
	Functional	**Narrative**	**None**
Introduction	58	4	12
Approach	24	47	0
Research design	10	18	0
Results	27	6	0
Discussion	39	18	0
Conclusion	68	1	0
Postlude	1	0	0
	227	94	12

Table 5. The RA section headings and their correspondence with the stages.

In the case of the introduction, the writers appear to prefer the functional term *introduction* in the majority of instances examined. In 12 cases, the writers do not provide a section heading for introductions; however, the section's position in the article, along with the content, enables this stage to be readily identified. Narrative titles are the least common label option. In the case of the approach stage, variation is more evident, and a narrative style tends to be preferred to a functional heading. The approach stage also shows the alternative words

framework, background, approach, review, foundations, and *research.*
These terms may appear pre-modified (by an adjective or a noun) or
post-modified (by a prepositional phrase). Examples include *model-
ing approach, theoretical background, theoretical framework, analytical framework, literature review, theoretical foundations, related
literature, tourism promotional literature, some notes on* (. . .), and
background to research. Evaluation is also signalled using the stance
adjectives *current* and *relevant,* as in *current research (on)* and *relevant previous research.* The word *methodology* is rarely used, save on
two occasions (*epistemology, methodology and terminology,* and *the
methodology*). In all of these sections, a review of the literature is presented.

The research design stage exhibits more cases of functional than
narrative labelling, as indicated in Table 2. Functional labels can be
divided into two groups. The first group contains nouns related to the
lexical field of procedure, namely *approach, method*(s), *methodology,
methodologies, design, models,* and *study.* The second group includes
words referring to evidence, such as *data, sources,* and *sample.* Items
from these sets may appear alone or in combination, as in the following occurrences: *methods and approach, research design and methods,* and *research approach and design.* In the *results* stage, functional
labels are the preferred option; these include the terms *results, findings,*
and *analysis/analyses.* Interestingly, two specimens contain the word
discussion in their formulation, *viz., results and discussion* and *findings
and discussion,* although the sections below these headings do not seem
to develop any kind of argumentation or justificatory claims.

In the discussion stage, a preference for narrative headings is evident. Functional headings are often signalled with the word *discussion*
alone or in combination with terms such as *wandering discussion* and
discussion of empirical findings. I have also identified other formu-
lations of this heading, including *analysis* and *situational features.*
Indeed, the discussion stage, despite being called *analysis,* does not
really introduce any analysis of the evidence.

Narrative headings in the conclusion stage are uncommon. This
stage is identified by the use of the word *conclusion* or by the use of
the present participle *concluding.* Other wordings include *implications, recommendations, remarks, thoughts, directions,* and *discussion.*

Except for the word *conclusion*, which may sometimes appear alone as a heading, the words mentioned above are presented in combination, as follows: *conclusions and implications, conclusions and recommendations, conclusion and discussion, by way of conclusion, concluding remarks, concluding thoughts, concluding discussion*, and *concluding remarks and directions for future study*. Another aspect of these functional headings is that they may be followed by a sequence in a more narrative style, as in the below instances:

(47) feeling like a child, feeling like a local (Gothie 2016).

(48) co-becoming with Country (Country 2017).

Finally, the postlude stage occurs only once in the entire corpus; this stage is explicitly called *postlude*, hence my decision to use this label. This stage has no concluding objective in the sense of summarizing the main contributions of the presented research. Instead, its function is to present new information that may reinforce or add a new perspective to the ongoing scholarly discussion and that emerged once the RA had been completed and accepted for publication. The added information does not seem to greatly modify the contents of the paper.

3.3.3 Frequent language templates per genre stage

In addition to section labels, the computer detection of certain linguistic expressions was very useful in validating my division of the texts into the functional stages listed in Table 5. For this purpose, I analysed the corpus per each stage to obtain 3-/4-gram types. The units detected were later checked using the key word in context functionality to obtain the concordances. Through the use of the n-gram queries, I was able to identify the presence and frequency of certain lexico-grammatical bundles that were clearly associated with each generic stage. Results were considered only if they appeared in several texts. In other words, despite their high frequency, lexico-grammatical patterns occurring only in one or two texts were rejected, as they may have indicated idiolectal significance and not genre significance. The corpus per genre stage was interrogated to obtain 4-gram expressions; where this did not yield significant results, I opted for 3-gram expressions.

The introduction stage contains expressions related to the presentation of the RA's purpose. Thus, I uncovered linguistic patterns with similar meanings, though they were otherwise realized using synonyms or related wordings, as in the following instances (the brackets indicate the optional use of variants separated by slanting bars):

The (ADJ) (aim/purpose/point) of (the/this article/study) is to INF:

(49) Hence, the aim of this study is to analyze the income elasticity of demand for nature-based tourism in Fulufjallet National Park (Fredman and Wikström 2018).

(50) The purpose of this study is to investigate tourists and local residents' perceptions of the Garden (Zhang and Ryan 2018).

(51) The focal point of this article is to offer reflexive insights into the methodological complexities that I as an insider researcher encountered and grappled with in situ (Vorobjovas-Pinta and Robards 2017).

This article's central purpose is to INF:

(52) This article's central purpose is to extend conceptualizations of existential authenticity by showcasing its scalar dimensions (Gillen 2016).

Alternatively, related expressions employing *have, focus on,* or *address* as processes may be used with the same function of presenting the article's topic or objectives:

The article focuses on (. . .) / The article has X goals (. . .) / This / Our article addresses (. . .) / The article presents (. . .)

(53) This article focuses on the two largest Catholic pilgrimages in Bosnia–Herzegovina (Valenta and Strabac 2016).

(54) Our article addresses a significant issue in tourism policy (Tzanelli and Korstanje 2016).

(55) This article presents three case studies where tourists have re-evaluated their role as visitors to tourist attractions (Zerva 2018).

In this stage, authors may also use a pattern in the present simple tense with a verb of ATTENTION, particularly the LOOK subtype, in which "the Perceiver [is] directing their attention to uncovering some information" (Dixon 2005: 217), or with a verb of SPEAKING of the DISCUSS subtype (Dixon 2005: 275), as in (56) and (57) below. Interestingly, the syntactic subject is filled either with an abstract entity or with the first-person plural referring to the authors of the RA. In the case of the DISCUSS type, the form *draw on* is used to introduce attribution for the method followed in the paper, as in (58). Other patterns in this stage may include a verb of the THINKING-f type, i.e., the CONCLUDE subtype (Dixon 2005: 488), or a verb of the SPEAKING type, particularly of the REPORT subtype, as in (59) below.

This article argues/explores/examines/shows/demonstrates (. . .)

(56) This article examines the adverse impacts of the replacement scenario on the Australian economy (Pham *et al.* 2018).

(In this article,) we argue/approach/explore/focus (ADV) on/(shall) examine/theorize/demonstrate (. . .)

(57) In this article, we focus on the carnivalesque at the English seaside from the perspective of tourism employees (Chapman and Light 2017).

This/The article draws on (. . .)

(58) This article draws on theory in the EM literature focusing on Morrish et al.'s (2010) (Fillis *et al.* 2017)

This article proposes/concludes (. . .)

(59) This article proposes an analysis of gendered representations of cruise tourism in promotional brochures produced by the cruise industry (Vanolo and Cattan 2017).

Another set of expressions is used to describe a paper's internal organization, as in (60) and (61). This explains the presence of verbs belonging to the REST-C type, specifically the PUT subtype, which refer "to causing something to be at rest at a Locus" (Dixon 2005: 106).

The article is structured (. . .) / The article has X (parts/sections) (. . .) / The article is organized (. . .)

(60) This article has seven sections including the previous introduction (Nilsson and Tesfahuney 2018).

(61) To explore the above-presented arguments, the article is organized as follows: the succeeding section introduces cruise tourism (Vanolo and Cattan 2017).

The approach stage presents a number of impersonal matrices containing effective or epistemic nuances (cf. Marín Arrese 2009b) that signal epistemological information along with the authors' stance toward the information that they either want to highlight, as in (62), (63) and (64), or provide epistemic justification for, as in (65), using the matrix *it is evident that* (. . .):

It is necessary to (. . .) (clarify/interact/differentiate/develop/use)

(62) However, at this point it is necessary to clarify the role of the carnivalesque in seaside resorts (Chapman and Light 2017).

It should be noted that ATTRIBUTION/EVIDENTIAL (. . .)

(63) [. . .] it should be noted that recent research by Ji et al. (2016) has revealed that Japanese domestic tourists in Niseko feel challenged by the different manners and customs of Chinese tourists in the area (Nelson and Matthews 2018).

(64) However, it should be noted some tourism scholars have explored these issues (Torabian and Mair 2017).

It is evident that (. . .)

(65) Nevertheless, from the extant literature it is evident that, over time, that significance has evolved in a manner similar to that of natural land environments (Jarratt and Sharpley 2017).

In this stage, there are language expressions referring either to a field of knowledge or to the association of fields or elements that merit consideration, as shown in these examples:

(. . .) in DET (ADJ) field of (inquiry/tourism/research) (. . .)

(66) The intersection of music and tourism has become a growing field of inquiry, encompassing a broad range of disciplines (Garcia 2016).

(. . .) the relationship between (FIELD/ITEM) and (FIELD/ITEM) (. . .)

(67) Some researchers have indicated a positive relationship between cultural inheritance and memorable experiences (Lee 2015).

As this stage is devoted to clarifying epistemological issues, a common bundle is the string *X (BE) defined (as/in)* to introduce meaning or description, as in (68). There are also cases in which forms of BE are omitted, as in (69).

(68) The term 'materialism' is defined as the importance that a consumer attaches to possessions that exist in the world (Correia *et al.* 2018).

(69) Semiotic resources, he argues, have semiotic potential, defined as a "potential for making meaning," that is theoretical and/or actual (Bratt 2018).

The research design stage is characterized by formulas including such method-related words as *analysis, data, interview,* and *method* as well as verbs associated with research processes, e.g., *conducted, designed, performed, focused on, carried out,* and *transcribed.* Evaluative stance is very rare in these formulas in this genre stage. The noun *data* may be modified by such adjectives as *anonymous, digital, empirical, formal, incomplete, initial, multimodal, qualitative, rich, statistical, visual,* and *ethnographic;* these account for any salient feature of the kind of evidence handled, as in (70) and (71). The template *(DET) interviews BE* normally presents the verb in the past tense, as in (72) and (73).

(...) analysis focuses on / has been performed (...)

(70) [...] and then the analysis has been performed by mobilizing a qualitative methodology (Country *et al.* 2017).

(...) (analysis of) (ADJ) data was/were (carried out/undertaken/ collected/conducted/employed/focused on) (...)

(71) In this approach, digital data are collected from traveler blogs, on-line tourist reviews [...] (Walter 2016).

(DET) interviews BE (carried out/completed/conducted/transcribed/undertaken) (...)

(72) All interviews were digitally recorded and fully transcribed (Jarratt and Sharpley 2017).

(73) The interviews were transcribed and interpreted using additional notes (Chapman and Light 2017).

In the results stage, common structures include the below expressions concerning the presentation of results and findings. The verbs associated with these expressions belong to both the DISCOVER and the SHOW semantic classes of verbs (Dixon 2005: 132–133), as in (74), (75), and (76). These expressions may be complete sentences, as in (77), matrices, as in (78), and/or adverbials, as in (79) and (80).

(DET) results/findings (reveal[ed]/show/[ed]/suggest/yield) (. . .)

(74) The results revealed that most of the tourists visited South Tainan Railway Station by motorcycle (48%) or by car (33%) (Lee 2015).

(DET) findings (suggest/demonstrate/show/lend support/illustrate/indicate/emphasize) (. . .)

(75) Our findings also show that providers advocate that through their offered (commodified) 'constructive' activities, gappers will not only help the host community but also get the opportunity to develop their personal skills (Hermann *et al.* 2017).

Results (are/were) (presented/summarized/plotted) (. . .)

(76) The results are summarized in Table 4 (Beritelli and Reinhold 2018).

(DET) results are presented in Table X

(77) [. . .] and the CFA results are presented in Table 2 (Park *et al.* 2016).

(Fig. X / Table X) shows that (. . .)

(78) Fig. 10 shows that before the age of 25, people with different educational attainment levels tended to have very similar total tourism expenditure (Lin *et al.* 2015).

As shown in Table X (. . .) (← Adverbial)

(79) As shown in Table 1, the estimates of *d* are significantly smaller than 1 in all cases, except for 3 Asia, in which the unit root null hypothesis cannot be rejected (Payne and Gil-Alana 2018).

As illustrated in (Table X / Fig. X) (...)

(80) As illustrated in Table 2, 46% of the respondents indicated they had some
level of knowledge about the GBR (Salvatierra and Walters 2017).

Analysis of the 3-gram structure reveals the presence of stance structures in the discussion stage, as in (81) and (82). In the case of (80),
the perspective shows epistemic stance-taking (Marín Arrese 2009b)
with the use of the modal *would* or the evidential *seem to/that* (Alonso-
Almeida and Carrió-Pastor 2015). As for (82), the template *it is important to + V* indicates effective meaning, i.e., the obligation or necessity
to carry out an action (Marín Arrese 2009b). In (82), the authors note
the need to consider a particular characteristic of certain rhythms and
how these relate in order to understand their point of view.

(...) NP (would seem / seems) (that/to) (...)

(81) However, the pilgrimage site does not seem to be often discussed or criticized by other groups in Bosnia–Herzegovina (Valenta and Strabac 2016).

(...) it is important to V (...)

(82) Thus, it is important to note that while some rhythms are more place-based
and others more mobile, they are not mutually exclusive (Rickly 2017b).

Formulas that refer to factuality, reformulation, attribution, addition, and
exemplification are also common in this stage as these instances show:

The fact that (...)

(83) [...] and the fact that the tour is conducted on foot rather than on a bus
highlights its bespoke nature (Johinke 2018).

(...) according to (DET) (ADJ/N) (terms/analysis/survey/the work
of) (...)

(84) According to Gilbert and Serge Trigano, the Club Med came to be accepted
among locals (Tchoukarine 2016).

(...) as (is/was/in) the case (of/in/with) (...)

(85) Trip leaders are less effective if they lose their legitimacy as was the case in this study (Liston-Heyes and Daley 2016).

(. . .) as well as (. . .)

(86) Pine and Gilmore (1998) note that festivals 'script distinctive experiences around enticing themes, as well as stage activities that captivate customers before, after, and while they shop' (p. 101) (Cashman 2017).

(. . .) in other words (. . .)

(87) In other words, destination image may originally be suggested by advertisements; however, it is not crystallized for a tourist until they engage with it personally (Lund *et al.* 2017).

Finally, the conclusion stage has certain structures in common with the introduction stage. The conclusion includes expressions containing verbs of the DISCUSS type, the CONCLUDE subtype of the THINKING-f type, and the REPORT subtype. As in the introduction stage, the use of the words *article, study,* and *research,* combined with the aforementioned types of verbs, is common in the conclusion stage. The pro-form *it* referring to any of these words is also an option:

(. . .) in this article I/we have (examined/presented/sought/developed/implied/deliberated) (. . .)

(88) In this article, we developed a frame analysis of the two largest pilgrimage sites in Bosnia–Herzegovina (Valenta and Strabac 2016).

(. . .) this article (provides/reveals/proposes/presents/shows/suggests/etc.) (. . .)

(89) This article proposes that geotourism resources at Mount Pinatubo are composed of its natural and cultural landscapes (Aquino *et al.* 2018).

(. . .) it could be (argued/addressed/assumed) (. . .)

(90) With this in mind, it could be argued that some aspects of what is referred to here as spirituality could also be accommodated into a discussion of seaside-related wellness (see Bell et al., 2015; White et al., 2013) (Jarratt and Sharpley 2017)

(. . .) it/we (was/has been/have) found (that/to) (. . .)

(91) [. . .] we have found that place pride, social cohesion, tourism benefits and sales improvements are expected and welcomed (Jiménez-Esquinas and Sánchez-Carretero 2018).

The verb *suggest* in the conclusion stage is used to introduce information that has been legitimized by the evidence previously discussed in the article. The concepts discussed in examples (92) and (93) exhibit different perspectivizing strategies. While the term *research* is the subject of the two matrices in (92), the pronoun *we* (referring to the authors) is deployed in (93). This indicates different degrees of authorial commitment and responsibility toward the propositions manifested in these instances. The use of the evidential expression *we have suggested* indicates explicitly shared responsibility. Even if it is obvious that all information provided in (92) and (93) stems from the work of the RA authors, the use of an explicitly designated conceptualizer (Langacker 2008: 438; Marín-Arrese 2009a: 238) appears as a strong indication of epistemic validity for the proposition following the matrix *we have suggested*. The use of *research* in (92) constitutes an opaque conceptualizer; as a result, the authors' responsibility here seems to be lesser than that in (93).

(we / research / our data / our findings / our analysis / this study/ research/work) (suggest(s) / have suggested / has suggested) (. . .)

(92) Research suggests that television campaigns (Eveland and Scheufule, 2000), as opposed to those delivered by print media, are the best way to reach this segment (Salvatierra and Walters 2017).

(93) [. . .] we have suggested that it also manifests from a deep rooted sense of duty linked to Japanese cultural identity (Nelson and Matthews 2018).

Finally, the formula *BE likely to + V* is used to indicate epistemic probability concerning the truth of the propositional content in deductive processes, as shown in (94). Following Marín-Arrese (2009a: 255), the subjective matrix *the revisit rate is likely to increase* is used to indicate implicit personal responsibility so that the authors are held responsible for the evaluation.

BE likely to + V

(94) Previous studies have indicated that if suppliers provide customers with unforgettable experiences, the revisit rate is likely to increase (Lee 2015).

As shown in this section, the lexico-grammatical bundles found in my corpus work to characterize each generic stage of the tourism RA. In addition, these formulas contribute to the arrangement of information while also signalling the authors' perspectives. In the remaining sections, I will comment on some other language aspects related to academic writing, with a specific focus on their variance per genre stage and their communicative functions.

3.3.4 Lexical density

Lexical density, i.e., "the number of lexical words per clause" (Halliday and Martin 1993: 83), has been reported to be higher in the written mode than the spoken mode, while scientific writing presents an even greater concentration of words per clause according to Halliday and Martin (1993: 84). Holtz (2011: 95) agrees that these texts are more "difficult to read" when lexical density values exceed 10 words per clause. In my texts, lexical density variance among the RA stages was detected using the LCA, and the results were processed to obtain the values of one-way ANOVA variance significance in Excel. The variance value was $F = 2.01$; moreover, as the probability value was 0.07 ($p < 0.05$) and the critical value was higher than F (i.e., $F\ crit = 2.23$), there was no variance significance among the RA stages in terms of the number of lexical words per clause.

Lexical density also measures informational density. As pointed out in Fang (2005: 338), this involves expanded noun phrases, as in the examples presented in (95). The use of premodification and nominalization was also evaluated in the detection of lexical density. Example (95) contains an extended noun phrase with the relativizer *which*, while (96) and (97) present instances of premodifying nouns and nominalization, respectively. These features contribute to creating what Biber and Gray (2011: 229) have labelled the "compressed" style of discourse in academic writing.

(95) Applying these positions to the DWEWT, displays of tourism-cultural capital which became converted into non-objectified, but individualised, symbolic capital simultaneously fulfil institutional requirements to uphold the brand values of *Doctor Who* and subsequently the symbolic capital of the BBC (Garner 2017).

(96) Emergency exit instructions (Barry 2017), family holiday meals (Schänzel and Lynch 2016), fan visitation opportunities (Lundberg *et al.* 2018), fieldwork documentation techniques (Barry 2017), home style accommodation (Nelson and Matthews 2018), hotel guest satisfaction (Garrigos-Simon *et al.* 2016), interaction partner relationship (Reichenberger 2017), media tourism contexts (Garner 2017), media on-screen production (Lundberg *et al.* 2018), memory discourse functions (Schäfer 2016), motorbike tourism company (Gillen 2016), museum visitor numbers (Ryan 2016), perception-imagination continuum (Gothie 2016), pilgrimage destination marketers (Park *et al.* 2016), pilgrimage travel experience (Park *et al.* 2016), pre-construction survey data (Jethro 2016).

(97) Embodiment (Nilsson and Tesfahuney 2018), obtainment (Chen 2018), visitation (Nelson and Matthews 2018), interactions (Walter 2016), construction (Cooke 2017), reflexivity (Vorobjovas-Pinta and Robards 2017), engagement (Jethro 2016), transformations (Gyimóthy 2018), generalizability (Salvatierra and Walters 2017), chanciness (Zerva 2018).

Variance analysis of the extended noun phrases revealed that this feature was statistically significant among the stages of the RAs, with an effect size of 0.027. In this case, the probability value was 0.015 ($p < 0.05$), with $F = 2.84$ and a critical value of 2.23. The use of premodifying nouns among stages in my RAs exhibited statistical significance with a probability value of 0.0001 and $F = 5.09$. The effect size value (i.e., $\omega = 0.066$) indicated a small effect. In terms of nominalizations, their use was statistically significant among the RA stages, as revealed by the one-way ANOVA. Here, the probability value was 0.001 ($p = 0.05$) with $F = 4.18$ and a critical value of 1.331. Variation in terms of noun premodification and nominalization is visually depicted in Figures 2 and 3.

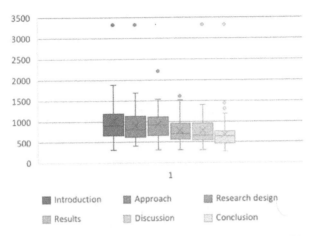

Figure 2. Premodifying nouns boxplot for all stages (95 percent confidence value).

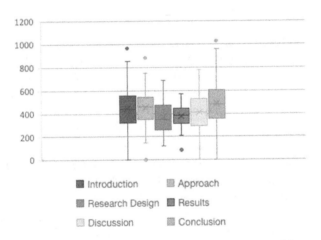

Figure 3. Nominalization boxplot for all stages (95 percent confidence value).

Adjectival premodification (e.g., *non-representational theories* [Chen 2018] and *endogenous variables* [Pham *et al.* 2018]) are also statistically significant, with a probability value of 0.006 ($p = 0.05$) with $F = 3.26$ and a critical value of 0.87. The variance among stages is plotted in Figure 4. The effect size value suggests a small effect ($\omega = 0.043$).

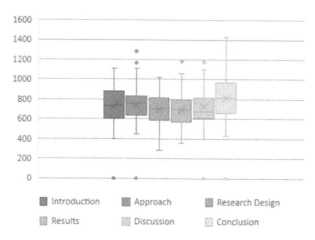

Figure 4. Adjectival premodification boxplot for all stages (95 percent confidence value).

3.3.5 *Syntactic complexity*

Measures of syntactic complexity are listed in Table 6. While these dimensions are typically used to obtain information concerning the level of proficiency of L2 learners, they are also helpful in supplying information about the underlying differences in syntactic elaboration and sophistication among the stages in the RAs I have examined. This is my area of interest in this section. I have therefore selected such variables at the clausal and phrasal levels, allowing for the addition of subordinated structures: clauses and dependent clauses per T-unit (C/T and DC/T, respectively), dependent clauses (DC/C), mean length of clause (MLC), verb phrases per T-unit (VP/T), complex nominals per clause (CN/C), and complex nominals per T-unit (CN/T).

		C/T	DC/T	DC/C	MLC	VP/T	CN/C	CN/T
Introduction	Mean	0.1294	0.2605	0.1997	49.5637	1.3489	1.5751	2.0420
	SD	0.0310	0.0536	0.0330	7.2488	0.0985	0.1349	0.1861
	Minimum	0.0606	0.1646	0.134	35.7351	1.1084	1.2167	1.4314
	Maximum	0.2063	0.4105	0.28	68.8714	1.5686	1.8936	2.5217
Approach	Mean	1.3310	0.2875	0.2147	51.0086	1.4054	1.6062	2.1370
	SD	0.0787	0.0591	0.0348	6.7211	0.1089	0.1263	0.2014
	Minimum	1.1744	0.1667	0.1389	39.9150	1.1935	1.3617	1.6847
	Maximum	1.5985	0.4712	0.3082	71.0216	1.7153	1.8987	2.7810
Research D.	Mean	1.3191	0.2753	0.2077	50.6871	1.3614	1.6051	2.1151
	SD	0.0777	0.0572	0.0360	7.0231	0.0887	0.1332	0.1931
	Minimum	1.1529	0.1698	0.1385	34.5212	1.1585	1.3657	1.7830
	Maximum	1.4937	0.4074	0.2778	66.1688	1.5556	1.8800	2.5921
Results	Mean	1.3146	0.2751	0.2084	52.8608	1.3839	1.6288	2.1399
	SD	0.0622	0.0490	0.0314	6.4428	0.0968	0.1362	0.1937
	Minimum	1.202	0.1685	0.1364	41.1159	1.2088	1.292	1.7805
	Maximum	1.4767	0.3535	0.2612	66.9802	1.6279	1.9681	2.7093
Discussion	Mean	1.3258	0.2753	0.2097	54.7198	1.3922	1.6374	2.1707
	SD	0.0633	0.0572	0.0355	8.5009	0.0938	0.1116	0.1786
	Minimum	1.1702	0.1698	0.1364	39.3176	1.2208	1.4018	1.7979
	Maximum	1.4390	0.4074	0.2956	77.2636	1.6021	1.9573	2.5730
Conclusion	Mean	1.3007	0.2587	0.1979	49.8062	1.3486	1.5740	2.0448
	SD	0.0635	0.0563	0.0360	7.7282	0.0891	0.1249	0.1613
	Minimum	1.1855	0.1587	0.1250	36.2761	1.0667	1.2816	1.6709
	Maximum	1.5312	0.4375	0.2857	68.9043	1.5833	1.8120	2.3833

Table 6. Measures of syntactic complexity (SD = standard deviation).

As this table shows, there are few differences in terms of the C/T exhibited in each stage. The introduction stages have the lowest mean (i.e., 0.1294), while the remainder have a mean of ca. 1.3 C/T. Another difference concerns the MLC, as the discussion stage actually over-scores the rest of the stages with a mean value of 54.7197. The discussion stage presents the highest scores in all dimensions, save for the VP/T. The use of longer structures in the discussion stage may result from the justificatory language used during the interpretative processes following the inspection of the results obtained in each article. Some examples are presented below.

(98) *Consequently*, the conception of 'on-screen dollying' reflects epistemo-logical elaborations on heritage interpretation (Uzzel, 1992), tourism

'world-making' processes in recreating social constructions (Hollinshead, 2009; Hollinshead *et al.*, 2009) and cinematic tourism network hermeneutics (Tzanelli, 2007, 2013, 2015) to better understand the role of film and television in making, remaking and unmaking places as tourist destinations (Lundberg et al. 2018).

(99) *For example*, some of the participants were reluctant to disclose information about their sexual encounters, or their deeper relationships with other resort guests; *however*, through the observation method, I observed some of these relationships *as* they were unfolding, and even witnessed romantic and sexual moments in their lives as they happened at the resort (Vorobjovas-Pinta and Robards 2017).

(100) *If* we take a more holistic approach that is grounded in human lived experience, *then* perhaps we will find that there is room for an enlarged care ethics to inform further patterns of environmentally responsible behaviour that can lead to emission reductions, *even if* a complete cessation of flying is not practical to achieve (Hales and Caton 2017).

These examples illustrate the use of discourse markers, such as *consequently* (98), *however* (99), *for example* (99), *as* (99), *if* (100), *then* (100), and *even if* (100), to introduce justificatory or exemplificatory claims, thereby providing lengthy descriptions of the subject matter. Complex noun phrases are commonly used to designate events in these examples with three or more units per phrase, as follows: (a) premodification (e.g., *tourism "world-making" processes* and *cinematic tourism network hermeneutics* in [98]); (b) noun post-modification with relative clauses with *that* (e.g., *holistic approach that is grounded in human lived experience* in [100]); and (c) subordinated complements introduced by prepositions (e.g., *in recreating social construction* in [98], *of film and television in making, remaking and unmaking places* in [98], *deeper relationships with other resort of these relationships* in [99], and *a complete cessation of flying* in [100]).

Evaluative language is also deployed in these excerpts in the form of *that*-matrices (e.g., *we will find that* [. . .] in [100]), epistemic expressions (e.g., *some of* in [99] and *perhaps* [100]), and dynamic expressions referring to a possible meaning in the sense of enablement (e.g., *can lead to* in [100]). In short, the expression of logical reasoning to account for the analysed data calls for a set of rhetorical strategies that make this reasoning process evident to readers. These strategies contribute to an elaborated and sophisticated syntax with an emphasis on argumentation.

3.4 Conclusion

This chapter has considered the rhetorical organization of RAs in the field of tourism. To my knowledge, this is the first attempt to describe the macrostructure of these RAs. My description relied on a SFL notion of genre to demonstrate that RAs show an identifiable textual structure, which may present with up to six functional stages. These stages may be introduced by conventionalized headings to signal what the stage's target is, or some of the headings may be more creative and narrative, relying on the contents of the RA.

I have also demonstrated the presence of certain language templates per genre stage. Many of these templates are in agreement with the objectives of the stage in which they appear. For instance, some of these structures, *viz.,* epistemic subjective matrices, have a clear evaluating scope to assess the writer's own findings and conclusions. The fact that academic writing needs to be precise, and sometimes even succinct, while it also has to create texture and reveal the authors' point of view translates into lexically and syntactically complex structures. I have also measured these in this chapter to learn that these structures may show variation per genre stage. I have seen that longer and more complex structures appear as the result of the use of justificatory language in stages that require major authorial involvement, e.g., the discussion stage.

4 Perspectivizing language in tourism-related research articles

4.1 Introduction

This chapter focuses on the perspectivizing strategies deployed in a corpus of RAs in the field of tourism per genre stage in order to detect variation among stages as well as to uncover additional communicative functions. As discussed in chapter 2, I have selected the following variables: tenses, modals, intensifiers, passive, conditionals, and *that*-complement clauses. They are common in the literature on stance-taking in academic and scientific texts. I use the corpus of RAs described in section 3.3.1 to assess their presence in tourism-related academic writing and to excerpt samples illustrating their meaning and use. References to the collected RAs are given in the appendix of this monograph.

As also explained in section 3.3.1, this study benefited from corpus linguistics tools; POS tagging was extremely useful for searching for and identifying these structures. For this purpose, I used CQL in order to interrogate my RA compilation with specific complex searches, e.g., evaluative *that*-clauses and passives. The data obtained was processed by statistical means to disclose substantial use in the RAs and significant variation in the use of these strategies per genre stage.

4.2 Tenses

Present tense exhibits statistical differences among stages, as does past tense, as seen in Table 7.

Variable	F	p (p < 0.05)	F crit	Effect size
Present tense	9.29	0.0024	2.23	0.114
Past	22.37	0.015	2.23	0.24

Table 7. Tenses: One-way ANOVA variance analysis measures.

Tables 8 and 9 offer the values for the mean, standard deviation, minimum, and maximum concerning the use of tenses per genre stage:

		Introduction	Approach	Research design
Present tense	Mean	491.3562	513.4057	346.1038
	SD	164.5723	136.8697	150.6219
	Minimum	0.0000	0.0000	83.8926
	Maximum	798.8981	767.8571	732.9843
Past tense	Mean	95.6572	115.2436	302.0268
	SD	96.4419	101.8916	180.1655
	Minimum	0.0000	0.0000	0.0000
	Maximum	361.7022	517.3427	748.0315

Table 8. Tenses: Mean, standard deviation (SD), minimum, and maximum values of the introduction, approach, and research design sections.

		Results	Discussion	Conclusion
Present tense	Mean	406.3652	489.1554	478.5994
	SD	178.2480	166.4005	171.9418
	Minimum	88.6228	0.0000	0.0000
	Maximum	788.3948	771.5582	883.1909
Past tense	Mean	269.1611	191.6780	123.7446
	SD	202.9018	153.6230	121.9332
	Minimum	19.0840	0.0000	0.0000
	Maximum	661.0779	673.8980	546.2185

Table 9. Tenses: Mean, standard deviation (SD), minimum, and maximum values of the results, discussion, and conclusion sections.

As shown in these tables, the present tense occurs fewer times in the research design stage and the results stage, with mean values of 346.1 and 406.36, respectively, while the past tense presents with higher means in these same sections of the RAs, with mean values of 302.02 and 269.16, respectively. However, the introductory and concluding sections do not have high values of lexical verb forms in the past tense, as they rarely report on previous events or actions.

In the introduction stage, the verbs occurring most frequently in the present tense are related to the notions of (a) contribution, e.g., *provide* and *include*, and (b) evaluative analysis and argumentation, e.g., *argue, focus,* and *draw (an idea, an assumption)*. These verbs contribute to setting the scene of research in the RAs, as illustrated in (101).

> (101) The investigation *focuses* on the ways the tourist experience is communicated in this particular context (Wang and Alasuutari 2017).

In this example, the verb *focus* introduces the object of study of the RA. This verb presents the conceptualizer implicitly to maximize the idea of objectivity in the RA, and consequently provides a reliable context for the RA's potential readers.

In the approach stage, the forms found in the present tense facilitate the objective presentation of argumentation and facts from the existing literature or from known touristic examples, as shown in (102) and (103), respectively. The verb *make* is used to either indicate an activity, whether physical or mental, as in (104), to indicate causative meaning, as in (105), or to indicate resultative meaning when *make* is combined with adjectives, as in (106). The use of the past tense with verbs referring to experiential and communicative events (e.g., *find, visit,* and *report*) is explained in the descriptive function of this stage to situate the study within a particular theoretical background and context of practice, as shown in (107).

> (102) Braithwaite (2003) *provides* insights into memories of hospitable occasions (Schänzel and Lynch 2016).

> (103) Notable exceptions *include* Disneyland and Disneyworld (Marling, 1997), Hobbiton in New Zealand [. . .] (Lundberg *et al.* 2018).

> (104) [. . .] visitors *make* a deliberate choice [. . .] (Weatherby and Vidon 2018).

(105) [. . .] when events *make* people remember together [. . .] (Zerva 2018).

(106) [. . .] all intended to *make* the sojourn there beneficial [. . .] (Straub 2016).

(107) A research study conducted by Singh and Best (2004) at the Hobbiton Movie Set *found* that the most important motivators were tourists attracted by the iconic attractions [. . .] (Zhang and Ryan 2018).

In line with the purpose of the research design stage, the verbs in the present tense are helpful for describing the way in which research on tourism is carried out. The past tense is used as a narrative strategy for providing a detailed description of the way in which data is obtained, as in (108).

(108) To locate research participants, we adopted both convenience and snow-ball sampling [. . .] We visited every hotel we found open and asked the receptionist or manager whether they employed anybody who had worked there during the socialist era (Dumbrăveanu *et al.* 2016).

In the results stage, the verbs in the present tense are related to the presentation and description of the findings obtained from the sample analysed, such as *show, reveal,* and *emphasize,* as exemplified in (109). The evidential forms (*suggest, seem to,* and *appear to*) are used to indicate inferential reasoning based on available evidence, as in (110), thus signalling perspective. Similar to the approach stage, this stage is strongly characterized by a narrative style with verbs in the past tense, as in (111). These are used to supply sufficient context to understand and explain the data obtained and to reinforce the idea of empirical and trustworthy research; consequently, this enhances the idea of reliability.

(109) The results reveal that the CR values of quality, price, risk and net value constructs as well as WTB are 0.85, 0.87, 0.93, 0.95 and 0.95, respectively (Shiu 2018).

(110) It was also found that Eastern China appears to be the key market for both group and non-group tourism (Lin *et al.* 2015).

(111) A non-significant w2 analysis revealed that past visitation had no influence on the time frame in which the respondent would return – w2 1/4 4.79 (2); p> 0.05 (Salvatierra and Walters 2017).

The discussion stage presents verbs in the present and past tenses, similar to those found in the results stage. Like the results stage, the discussion stage incorporates verbs related to experiential processes to signal stance; this is explained by the authors' intention to describe instances of their data and substantiate their claims, as exemplified in the following examples (with *involve* in the present tense and *feel* in the past tense):

(112) The use of brands involves a vastly complex network of relations deter-mining credibility and social capital (Nuenen 2016).

(113) The following climber described how she felt like an outsider for the first couple of months of camping at Miguel's (Rickly 2017a).

The use of the present tense with communicative and cognitive verbs clearly allows for the presentation of a point of view, as in the below examples with *argue* and *suggest*, construed with explicit conceptual-izers showing a subjective and intersubjective stance, respectively. In both cases, the authors are solely responsible for the information they have cognitively elaborated.

(114) [. . .] but I argue that punk has always traded commercially on its outsider cachet (Johinke 2018).

(115) In presenting the framework, we suggest that on-screen tourism place-making involves the transformation of heritage (Lundberg *et al.* 2018).

The conclusion stage contains verbs in the present tense connected with the ideas of authorial perspective and contribution to general knowledge (e.g., *suggest, argue, demonstrate,* and *emphasize*). This is in accordance with the functions performed by the conclusion stage in the RA, namely to convey the main contributions of a paper and indi-cate the authors' positioning with respect to observed data and within existing knowledge. The following examples are illustrative:

(116) In conclusion, this study demonstrates that image perceptions towards nat-ural tourism icons such as the GBR are likely to change [. . .] (Salvatierra and Walters 2017).

(117) The latter seems to direct the search for uniqueness in past experiences [. . .] This indicates that tourists' relation to a place in retrospect represents a second chance to make the past visit significant to them and to others (Zerva 2018).

In (116), the form *demonstrates* in the present tense is used to clearly convey the epistemic validity of the propositional content; this verb also underlines the concept of successful research and trustworthiness. In the case of (117), the evidential predicates *The latter seems to* (...) and *This indicates* (...) suggest opaque personal responsibility, and the speakers are the conceptualizers in this context, as Marín-Arrese (2009a: 255–256) points out.

4.3 Modals

Modal verbs in the RAs include the central forms *can, could, may, might, must, will, would, shall,* and *should.* Table 10 shows that the use of modals among genre stages is statistically significant. Standard deviation, minimum, and maximum values are presented in Tables 11 and 12.

Variable	F	p (p < 0.05)	F crit	Effect size
Modals	13.49	0.049	2.23	0.158

Table 10. Modals: One-way ANOVA variance analysis measures.

	Introduction	**Approach**	**Research design**
Mean	52.7038	61.7577	42.4267
SD	41.4829	29.6099	31.6617
Minimum	0.0000	0.0000	0.0000
Maximum	168.5393	126.4980	167.9389

Table 11. Modals: Mean, standard deviation (SD), minimum, and maximum values of the introduction, approach, and research design sections.

	Results	**Discussion**	**Conclusion**
Mean	62.2577	78.6047	97.1391
SD	32.9852	40.7417	64.5008
Minimum	0.0000	0.0000	0.0000
Maximum	126.9130	236.6127	402.6845

Table 12. Modals: Mean, standard deviation (SD), minimum, and maximum values of the results, discussion, and conclusion sections.

As Table 12 shows, higher mean values are found for the discussion and the conclusion stages. This could be explained with reference to the perspectivizing force of modal verbs, the use of which is an effective way to develop argumentation in these two stages. In my corpus, the meanings of modal verbs are probability, possibility, obligation, and necessity. These meanings may fulfil different pragmatic functions, e.g., to indicate politeness and provide advice, as in the following instances:

(118) A cross-cultural study that compares countries or ethnicities *could* enhance the understanding of cultural effects (Park *et al.* 2016).

(119) Accordingly, efficiency *may* explain the reverse direction of causality (Mendieta-Peñalver *et al.* 2018).

(120) It *should* also be noted that significant attention has been paid to the physical health benefits of natural, green spaces, although this literature is beyond the scope of this article (Jarratt and Sharpley 2017).

(121) To achieve better performance, travel agencies *must* develop an IPO strategy by raising barriers to competition in order to strengthen their branding value and align their firm organization culture (Huang and Chang 2018).

The form *could* in (118) indicates epistemic meaning, evaluating the chances that a cross-cultural study could clarify how knowing cultural effects might impact tourists' anticipations. Possibility, in the sense of enablement (i.e., dynamic possibility), is conveyed in the form *may* in (119); this modal suggests the potentiality of the notion of efficiency to justify the phenomenon of reverse causality. Both *could* and *may* are used as negative politeness strategies (Brown and Levinson 1987) to avoid imposition on the readers. The form *should* in (120) conveys deontic obligation in the sense of advisability, as it directs the readers' attention toward a specific body of literature in the domain of tourism. This imposition may work to demonstrate the authority of the authors in this research field to show how the research may help the tourism industry become more competitive. In the case of *must* in (121), this modal entails necessity deontic meaning in this context, as the action framed by the modal is needed to achieve better market results. In this way, the authors also signal their expertise in the field by openly committing to the truth of the propositional content.

As noted earlier, modal verbs evidently appear in all genre stages of the RA, although they show variation according to the meanings they represent in discourse. In many ways, these meanings have a certain correspondence with the function of the stage they appear in, as in (122):

> (122) Embodied tourism experiences *can*, therefore, serve as a trigger for reflecting on stereotypes or rethinking relationships (Gibson, 2009). Indeed, tourism *may* be a powerful medium for learning that transforms tourists' habits of thought and ways of engaging with the world. For BCE, hosting visitors is a way of sharing knowledge and reasserting control over how Yolŋu knowledge is communicated. Tourism forms one component of a long history of Yolŋu empowerment and resistance to colonialism through a wide range of areas including art, politics and science (Morphy, 1983; Watson-Verran and Chambers, 1989; Watson-Verran and White, 1993; Williams, 1986). (Country *et al.* 2017)

In this excerpt, there are two modals, namely *can* and *may*, which encode dynamic meanings. The form *can* indicates the suitability of tourist experiences to promote reflection that can change the actual behaviour of tourists. This is also the case of *may* in the subsequent sentence. This form reinforces the idea of the potentiality of tourism to grant empirical information that can be useful for the same function. The logical discourse unit *therefore* and the evidential *indeed* contribute to the elaboration of meaning based on the actual properties of tourism as a field to offer substantial information for research. Additionally, tourism has social, cultural, and political implications that require analysis. The function of these two modals might be to mitigate the effects of potential face threat brought about by the expression of factuality. The dynamic modals in this context refer to the same existing reality without clearly imposing the authors' point of view. In (123), the form *may* is used to show inferential communication:

(123) As a consequence of this policy change, the measurement of tourist arrivals reported by the National Travel and Tourism Office changed as well. Thus, the revised data on tourist arrivals from September 2001 forward is not directly comparable to previous arrival counts. Given this change in policy, this research note examines the degree of persistence in US tourist arrivals, over the entire period in which data are available from January 1996 to August 2016 (which encompasses the period for the Cunado et al. (2008a) study) along with two subperiods; January 1996 to August 2001 and September 2001 to August 2016, in order to determine the extent to which the change in data measurement associated with the requirement of the INS I-94 entry form *may* have impacted the degree of persistence. (Payne and Gil-Alana 2018)

This excerpt at the end shows a hypothesis introduced by the epistemic necessity modal, which is integrated within an argumentative passage with discourse logical markers e.g., *as a consequence* and *thus*. Inferential meaning follows from the use of this modal form with the perfective, i.e., *may have impacted*, resulting in what Boye and Harder (2009) refer to as evidential substance. This use of the modal allows the author to present the research question in the form of an inference based on the knowledge that is previously given in this same text. This is a common way of using modality in the introductions of research papers, as it helps both to introduce new ideas and to attenuate their illocutionary force.

The use of *shall* and *will* in (124) with deontic nuances is justified by one of the functions introductions have, i.e., to show the organization and contents of a research paper:

(124) We *shall* first provide a theoretical discussion that acknowledges how destinations are mobile entities and in a constant state of becoming. We then *will* provide an historical overview of the destination image of Iceland suggesting that the changing image of Iceland is crucial to understanding the present state of affairs. This *will* be done using a postcolonial approach recognising that postcolonial power relations between a hegemonic centre and its periphery are implicated in the production of travel representations that inspire destination image (Pratt, 2007). We *will* then use a semiotic approach to analyse recent Icelandair advertising materials, emphasising, in particular, the recent MyStopover campaign. (Lund *et al.* 2017)

The use of *shall* and *will* in this sample indicates the volitional meaning of covering certain aspects in the text in a chronological order, as also evinced by the use of *first* and *when*. Actually, these two modal verbs constitute one of the prerequisites of the four promissory acts this paragraph develops. All the events described are planned to occur in the future, altogether describing the internal organization of the paper.

The approach stage of a research paper includes central modal verbs with clear indexical functions. This is the case of the deontic *will* and *must* in the following instance:

> (125) Commenting on the geographic limitations of the earth, Kant (1996 [1795]) observed that humans *will* necessarily always come into contact with one another and so *must* also recognize a "cosmopolitan right" to travel as well as abide by a law of "universal hospitality" which allows for movement of individuals without hostility and obliges travelers to be neither exploitive or oppressive (see also Barnett, 2005; Derrida, 1999; Molz and Gibson, 2007; Popke, 2007) [. . .] An ethics of hospitality is not simply unconditional, but as Barnett (2005) explains, it is an ethics of tolerance that challenges the host's mastery of their own space. Thus, "tolerance is extended to a guest whose identity is already attributed" (Barnett, 2005: 11) through the temporalities of their arrival (Derrida, 2002) as a visitation or an invitation, a surprise or an expectation. For hospitality to be ethical, differences in the temporalities of the encounter matter and the hospitable act *must* acknowledge the identity of the guest and not associate them with a generic "Other" (Barnett, 2005). (Rickly 2017a)

The form *will* indicates necessity, and this is supported by the neighbouring adverbial *necessarily always*. This foregrounds the presence of *must* preceded by the logical result operator *so* to indicate that the event modulated by *must* follows from logical reasoning. The last instance of *must* in (125) clearly establishes the authoritative voice of the authors in the conditionality revealed by the structure *for* (. . .) *to*-infinitive given in the topical position. The whole claim is justified with evidentiary attribution material (cf. Hyland 2005) given parenthetically in the final clausal position as a way to reduce likely future counter-argumentation. In (126), the forms *may* and *can* have a clear dynamic sense:

> (126) Although it *may* subsequently be assumed that interactions with other tourists – especially other backpackers – constitute a large part of the social aspect of backpacking, little research has actually been conducted to further examine this issue and explore the role that social encounters

play for backpackers in the context of increasingly institutionalized tourism infrastructure [. . .] Considering the strength of the social motive in the backpacker experience, further research is required to address this aspect in more detail to understand not only the extent of its relevance but also the ways in which social interactions *can* impact, change or even produce individual experiences and their interconnectedness with the commonly overarching goals that underlie this particular travel form. (Reichenberger 2017)

The modal *may* modulates the evidential cognitive expression *be assumed that* in order to tentatively indicate the possibility laid down by contextual factors (hence its dynamic sense) to meet the conclusion following the *that*-matrix. The intersubjective nature of the complete matrix relies on the use of an opaque conceptualizer to minimize potential future conflicts for the claim. Something along this line is gained with *can* in the final sentence, as the authors want to make clear that the role social interaction plays on individual experiences will largely depend on contextual enabling conditions that are expected to show. Altogether these dynamic senses come to serve as a justification of the authors' research interests in their paper. Epistemic cases of *may* also appear in the approach stage, as in (127):

(127) Tourists *may* feel dissatisfied if experiences do not match prior perception and feel satisfied if the actual experience matches the hyper-real expectation (Carl et al., 2007) (Zhang and Ryan 2018).

In this instance, the epistemic modal verb is deployed to signal probability in the apodosis with a clear mitigating function. The event is presented as a hypothetical scenario provided the information in the protasis is fulfilled. The other occasions of modals in the apodoses in this section present the form *can* with a dynamic sense. The modal verb *can* captures the necessity of enabling circumstances for the event to happen, these being described in the protasis, as in the below example:

(128) The following hypothesis *can* be proposed and tested if quality can also relieve the risk perception in spa hotel choice (Shiu 2018).

In the case of the research design stage, dynamic modality is also used to suggest potential circumstances and possibilities, as shown in (129):

(129) At the same time, it has to be stressed that the performing of the content
analysis produced meaningful outcomes; first, it allowed us to perform
a first exploration of the data, providing inputs for successive qualitative elaborations; second, testing the limits of this methodology *can* be
considered, somehow, a research outcome in its own respect; ultimately,
considerations about the statistical distribution of specific visual contents
allowed us to better perceive the recurrence of particular elements and
the insistence on particular frames and tropes in promotional discourses.
(Vanolo and Cattan 2017)

The presence of *can* in this instance has the function of softening the
factual claim made, i.e., *can be considered a research outcome*. The
dynamic condition of this verb suggests enablement according to a
given set of contextual conditioning factors previously assessed by the
speakers. In the following excerpt, dynamic modal *can* has an obvious
justificatory target:

(130) Travel bloggers were identified as 'key [information] streams' (Lugosi
et al., 2012): as people on the move, the travel bloggers represented
'information-rich cases' (Patton, 2015) to investigate tourists' food and
drink-related experiences. More specifically, blog analysis *can* provide
personalised, informant-driven insights into experiences in and across
multiple destinations and consumption places (cf. Ji et al., 2016; Mag-
nini et al., 2011; Wang et al., 2016; Wu and Pearce, 2016). (Bezzola and
Lugosi 2018)

The form *can* is used here to assert a potential characteristic and value
of analysing blogs for the research carried out. As such, modulation
facilitates the introduction of the exact reason, without imposition,
why a particular methodology has been selected. The dynamic sense
of *can* is reinforced with the attribution material given at the end in the
sentence. Use of the dynamic modal meaning to suggest a particular
enabling feature of blog analysis is only understood in a context where
this device is deployed as a negative politeness strategy. The reader is,
therefore, provided with a reason that substantiates the claim made.
The results section provides examples of modals that are deployed to
imply factuality:

(131) Regarding the tourist experience, object-related authenticity *can* be seen
in the data in how sights are morally valued in the online discussions. One
evident instantiation is the use of linguistic markers showing how these

sights and attractions are favoured strongly in the group members' communication. (Wang and Alasuutari 2017)

The dynamic modal *can* in (131) signals the actual possibility to evaluate evidence from the results obtained, and this sense does not really need to be actualized, as it is based on a feature already exhibited in the authors' justification for the claim, i.e., *in how sights are morally valued in the online discussions*. From a pragmatic point of view, *can* operates as a negative politeness strategy. This is explained by the authors' involvement in the elaboration of meaning, which is evident in the use of this modal verb. The deictic force behind *can* shows the authors' willingness to substantiate their statement by ostensibly pointing out the correct selection of the contextual variables leading to their statement. The absence of the modal in such an utterance is possible, but this utterance would lose the authors' intention to show their implication; consequently, the polite behaviour would also be lost. Deontic meaning is used in this stage to indicate commitment to the research being presented:

(132) However, the younger groups give a good indication as to the future trend of potential changes to long-haul holidaying as this segment matures in age. Around one-third of the respondents overall acknowledge flying 'very often' with LCCs for leisure purpose, while a further 29% stated 'sometimes', while the majority (40%) revealed that they previously travelled with a LCC just a 'few times'. Finally, 56% of the sample confirms that their travels are accompanied by family, 42% by friends and just 2% travelling alone. These characteristics *will* be used as independent variables to carry out the analysis. (Martín Rodríguez and O'Connell 2018)

In this instance, the deontic *will* serves to indicate an evaluation of the findings described earlier in the paragraph. The promissory act shown in the utterance that includes *will* discloses its importance for the analysis to be carried out.

In the discussion stage, modal forms present a variety of meanings connected with the argumentative style associated with this stage of the genre. In (133), the modal form *can* entails dynamic possibility in the sense that the realization of the events described depends on the potentiality of the entities and not on chance:

(133) However, each of these studies overlooks detailed analysis of the per-
       formed role of tour guides within official media tourism contexts and the
       insights that applying a Bourdieusian perspective *can* provide. For exam-
       ple, given the myriad roles that tour guiding involves, what forms of cap-
       ital are demonstrated on official tours? Additionally, as symbolic capital
       can be accrued at institutional or individual levels, how is this constructed
       and ultimately located (e.g. institutionally or individually) on a tour like
       the DWEWT? (Garner 2017)

The pragmatic function is to convey politeness, thus avoiding potential threats. This is done by evincing the fact that the argumentation is
based on factual, probably attested, values of these entities to impinge
the realization of the actions modulated. Even if the use of these two
instances of *can* in (133) clearly report on their own set of potentialities
and dispositions, the last instance seems to lie semantically halfway
between dynamic and event possibility, as a sense based on permission
that can be contextually grasped in *symbolic capital can be accrued at
institutional or individual levels*. This meaning is not inconsistent with
the pragmatic function attributed to *can* as a dynamic modal. However, its categorization as a deontic modal reflects an intersubjective
functioning, which reduces the degree of the authors' contribution in
the elaboration of the information they are presenting. In this case, the
authors minimize the risk of potential face threat.

The below instance contains samples of deontic modals to convey
meanings pertaining to effective modality in the sense in Langacker
(2010):

(134) The data mining results *must* be presented in a readable form to the hotel
       management. It is important to give feedback to the management on the
       identified patterns pointing out when and where customer dissatisfaction
       has been detected, so management *can* implement corrective actions to
       solve the problem. As commented before in the evaluation phase, these
       results *can* be compared with other hotels in the chain to have more
       feedback about the services available and their impact on the market.
       (Garrigos-Simon *et al.* 2016)

Along with the conclusion stage, the discussion stage is peppered with
advice concerning the application of empirical research to the tourism
industry. In this context, the deontic *must* in the example above reinforces this idea of necessity, even if its use comes across as authoritative

and imposing. The presence of the matrix *It is important to* given as subsequent co-text might try to diminish the effects of this imposition, as it introduces a justification of the authors' viewpoint. In this sense, the modal form *can* that occurs later signals the dynamic possibility that follows from earlier argumentation. The use of the logical particle *so* also reinforces this dynamic interpretation, as the suggested course of action allows for the implementation of corrective measures. The last form of *can* in (134) also indicates dynamic possibility in that the existence of the results obtained enables the comparison of the data to be made so that *can* appears to mean *to be in a position to*. The following excerpt also shows a combination of dynamic, deontic, and epistemic modalities in the elaboration of meaning:

(135) Public space is not, in itself, a space for emancipation, but it *can* be disruptive, active and generative (Massey, 1999) [...] The way FP occupies public spaces is a type of challenge that also *must* be understood in relation to the facts of 15M. After 2011, activists relocated on a smaller local scale, working on more specific conflicts at a micro level (Walliser, 2013) [...] Although the seemingly festive atmosphere *might* suggest simplicity in the event, it actually politicises the space through strategies and tactics (De Certeau, 1984). (Bruttomesso 2018)

The form *can* entails dynamic possibility, as it reflects a capacity of the entity, i.e., *public space*, to be something else. This possibility is later exemplified with how *FP occupies public spaces*. In this context, *must* expresses deontic necessity, which could be seen as a kind of imposition on readers to guide them through the authors' line of argument. Unlike other cases of deontic forms described here, I have not been able to identify any attenuating linguistic strategy surrounding *must*. On the contrary, the authors' view is further and intentionally reinforced in the closing lines. While the epistemic attenuating devices *apparently* and *could* are given to imagine a hypothetical scenario, the evidential form *actually* that complements the present tense verb *politicize* makes it clear that the authors are pursuing an unambiguous authoritative stance.

The case of *may* in (136) is also consistent with the argumentative nature of the discussion stage:

(136) Despite its different rationale, the ROH project's desire for insularity also remained a key mechanism in creating the feeling of familiarity. First,

the collective nature and relative remoteness of the Praha resort *may have facilitated* feelings of home, for contacts with Yugoslavs were minimal. (Tchoukarine 2016)

The form *can* is used to indicate epistemic necessity. The combination of this modal with the lexical verb in the perfective is evidence of the authors' inferential mode of knowledge. The fact that the authors share with the readers the way of elaborating meaning suggests absence of imposition. The authors presuppose reasoning based on their available evidence but present it as a conjectural expected probability rather than as a fact.

In the conclusion stage, modal verbs are frequently associated with the functions of this part of the article, as shown in the following excerpt:

(137) The result obtained in the empirical section is conditioned by the existing data that represent the main limitation of the analysis carried out. An increase in the size of the sample and its variability *could* be expected by incorporating hotel chains from other countries, either consolidated or emerging as a result of the strong process of globalization experienced by the hotel industry in recent years. To continue with this line of research, we suggest to study whether the presence of foreign direct investment in the hotel industry gives rise to an increase in the competitiveness of the destination of the investment, due to knowledge transfer. Accordingly, efficiency *may* explain the reverse direction of causality. (Mendieta Peñalver 2018)

In (137), the verb *could* is used to suggest epistemic probability to evaluate the benefits of increasing the actual sample of the study. On a tangential note, it also assesses the authors' conclusions in relation to existing knowledge. The claim in which *could* is embedded has an indexical force that seeks to justify the claim made in the last lines of this paragraph, which is introduced by the adverbial *accordingly* in the topical position. This device obviously comes to introduce the result of the authors' reasoning process, and it can therefore be classified within the result/inference category of adverbials following Biber et al. (2021: 869). The result is expressed with the dynamic modal verb *may* to suggest the capacity of the notion of efficiency to account for the phenomenon under study.

The following excerpt focuses on highlighting the contributions of the research presented, and an array of modal meanings, mostly dynamic and deontic, are combined to unfold the argumentation:

(138) This research has shown that the drive and ambition of an innovative owner/manager can cut through these obstacles in order to develop a new venture. Many vacation marketers have an intuitive inclination to exploit their competencies in imagination, vision and non-standard solution finding, but others *will* need encouragement to step away from long-held beliefs. Many in the latter category have the potential to develop their businesses further, but they need to step out of their comfort zones in order to realize their full potential. Those tourism firms with entrepreneurial flair *will* always be inclined to take risks (see Ross, 2003), but others *can* be encouraged to move from being reactive. Innovative tourism firms *may* have a unique product and are shaped by equally innovative owner/managers who *can* exploit their own brands and identities in the marketplace in order to create new customers and, hence, new demand. (Fillis *et al.* 2017)

The first form *will* in (138) is deontic, as it reports on the necessity of a group of marketers fulfilling the action described. This idea of necessity is reinforced by the co-textual lexical elements *need (to)*. The second case of *will* shows dynamic necessity, and the authors rely on a common feature of the tourism firms mentioned to predict their reaction while also finding support for their claim. The authors' perspective changes to incorporate suggestions framed by a set of dynamic modal verbs entailing possibility, namely *can* and *may*. Although these modals do not have an intrinsic attenuation function, they allow for the gentle presentation of these recommendations, thus using a less authoritative tone than deontic modals. This is achieved through a focus on the capabilities and inclinations shown by individuals or other entities to complete an action. This explains, for instance, the use of *can* in "owners/managers who can exploit their own brands," as the authors rely on the owners'/managers' capacities to leverage their brands.

In contrast to this moderate presentation of recommendations for business improvement based on empirical research, the conclusion stage contains deontic modals to introduce the set of procedures needed either to improve the tourism market or to conduct further research. This is clear in the following instance:

(139) Aside from the need to further confirm these results through examining different cultures, some suggestions *should* be proposed to enhance the shopping for luxury items in HK, at least among Chinese tourists. Last but not the least, the geographic scope of this article limits the generalizability of its findings. Therefore, future research *may* encompass tourists from different cultures. Furthermore, the effect of socio-demographic characteristics and cultural values *should* also be considered. (Correia *et al.* 2018).

The form *should* in (139) reveals the authoritative position of the authors, and so, far from mitigating their claims, they express advice openly thanks to the evidence they obtain from their research. The dynamic modal *may* has a clear evaluative dimension in this context and therefore underlines the strengths of the research. Indeed, the current study seems to have facilitated the idea given in this example that "future research *may* encompass tourists from different cultures." Thus, the form *can* nicely captures this enabling sense to suggest that the conditions are now in place to undertake new research covering additional recommended variables.

4.4 Boosters and downtoners

Boosters and downtoners in tourism-related papers provide information concerning writers' potential degrees of commitment to certainty. The use of boosters in RA stages is statistically significant, as disclosed in Table 13. The mean, standard deviation, minimum, and maximum values are presented in Tables 14 and 15.

Variable	F	p (p < 0.05)	F crit	Effect size
Boosters	4.8	0.00029	2.23	0.062
Downtoners	2.29	0.045	2.23	0.03

Table 13. Boosters and downtoners: One-way ANOVA variance analysis measures.

		Introduction	Approach	Research design
Boosters	Mean	6.3513	10.1448	12.5523
	SD	11.0980	11.1814	16.7786
	Minimum	0.0000	0.0000	0.0000
	Maximum	51.6351	68.8705	85.8369
Downtoners	Mean	10.9345	16.7563	13.1731
	SD	16.4385	14.4543	12.7427
	Minimum	0.0000	0.0000	0.0000
	Maximum	62.6304	75.7575	55.3505

Table 14. Boosters and downtoners: Mean, standard deviation (SD), minimum, and maximum values in the introduction, approach, and research design sections.

		Results	Discussion	Conclusion
Boosters	Mean	18.2469	14.2829	11.5125
	SD	11.5602	11.1980	15.7349
	Minimum	0.0000	0.0000	0.0000
	Maximum	42.4448	45.3857	83.6820
Downtoners	Mean	20.2275	14.9719	17.0251
	SD	16.7014	11.2019	20.9160
	Minimum	0.0000	0.0000	0.0000
	Maximum	79.7342	48.2625	100.0000

Table 15. Boosters and downtoners: Mean, standard deviation (SD), minimum, and maximum values in the sections of results, discussion, and conclusion.

Examples of boosters in my corpus include the items *highly, completely, fully, thoroughly, greatly, very, perfectly, extremely,* and *strongly* among others. Some representative downtoners are *almost, scarcely, slightly, nearly, only,* and *partially*. The highest means of downtoners and boosters occur in the results stage. The emphasizing effect of boosters (Fest 2015) and the scalar nature of downtoners (Quirk et al. 1985: 445) mark the stance of the authors regarding the findings. With boosters, the authors may want both to intentionally raise attention to specific facts and to evaluate them. The boosters also indicate viewpoint, but the information is presented in a less assertive way, as shown in the below instances from the results stage:

(140)   However, we also observe that the seasonal AR coefficient is *extremely* large in practically all cases, implying that seasonality matters (Payne and Gil-Alana 2018).

(141)   However, 63% favoured purchasing travel insurance while *almost* half of the travellers would book the airport parking (Martín Rodríguez and O'Connell 2018).

The use of *extremely* in (140) is intended to signal the authors' assessment of the seasonal AR coefficient based on their empirical study, as evidenced by the use of the experiential evidential *we also observe that*. The latter softens, somehow, the use of the intensification in the subordinated clause. The example in (141) contains the downtoner *almost* and is used in an epistemic sense to achieve a mitigating effect in the presentation of the facts (Hyland 2005; Caffi 2007; Carrió-Pastor 2017). This function seems to prevail in the corpus since, as shown in Tables 14 and 15, downtoners have a higher mean number of occurrences than boosters in the corpus in all stages of the RA.

As to the use of boosters and downtoners concerning each generic stage, there is a tendency to use the same type of device in almost every section in the medial position, although there are a few genre stages showing a wider variety of them. In the case of the introduction, the boosters identified are *completely, fully, greatly, highly, strongly, very,* and *thoroughly*. The downtoners are *almost, merely, nearly, partially, partly, slightly,* and *somewhat*, although the form *only* is by far the most common. The following excerpts illustrate the uses of these devices in the introduction:

(142)   In order to analyse how destination branding, heritage propertisation and corporate culture are permeating two coastal villages in Galicia (Spain), we are not only looking at the 'host–guest' relations that have been *thoroughly* analysed in anthropology, tourism studies and destination branding literature (Cohen, 1988; Urry, 1990). (Jiménez-Esquinas and Sánchez-Carretero 2018)

(143)   Grimwood argued that tourism in the Arctic is *strongly* associated with Western-centric perceptions of modernity, whereas the voices of Aboriginal inhabitants remain silenced (Vorobjovas-Pinta and Robards 2017).

(144)   Travel to view unique and exotic landforms is not new, yet a concept to better understand this phenomenon has *only* been recently developed (Aquino *et al.* 2018).

The boosters *thoroughly* and *strongly* in (142) and (143), respectively, are used with different ends in mind. The form *thoroughly* clearly informs about the authors' consideration of a particular issue in the field in relation to the existing literature. In this sense, they show their expertise and good command of the topic to justify their research objectives. The adverbial *patently* signals the authors' position toward the information presented. This is not the case of *strongly* in (143), as it apparently refers to the third-party perspective attributed in the evaluative matrix *Grimwood argue that* (...). The downtoner *only* in (144) has a justificatory function to emphasize the need for the research proposed, as the subject matter has been somewhat neglected in the literature although research on the topic has a long tradition.

Boosters in the approach stage share some of the forms with those in the introduction, but this list has some other forms: *altogether, completely, entirely, extremely, finally, greatly, highly, perfectly, strongly,* and *very*. The most common is *highly*. Some examples are given below:

(145) Some critics of the semiotic model might argue that the quality of the analysis is almost *entirely* dependent upon the expertise and the cultural positioning of the analyst (reader) undertaking the study. Others may argue that semiotic readings are largely qualitative impressions. (White 2018)

(146) Although the extant literature fails to *fully* explain the nature of the underlying social contingencies that produce chance meetings, it offers three conclusions that inform the subsequent conceptualization of chance meetings: First, personal contacts exert direct as well as indirect influence on travel decisions and behavior. Second, certain social obligations produce planned or predetermined meetings. Finally, social contingencies can be connected to a likelihood of co-presence even for seemingly random encounters. (Beritelli and Reinhold 2018)

(147) With the above theoretical foundations, it can be implied that geotourism is about the construction of tourism spaces through the commodification of landscapes, geosites and geoheritage, for tourist appreciation and learning. Various researchers have *highly* focused their attention on this aspect of geotourism, as evident in the multitude of studies on the identification, quantification and scientific valuation of geosites (Brilha, 2016; Reynard et al., 2016); assessment of landscapes for geotourism development (Dos Santos et al., 2016; Višnić et al., 2016); and the creation of products (e.g. geotours) for geotourist enjoyment (Norrish et al., 2014; Santangelo et al., 2015). (Aquino *et al.* 2018)

All of these excerpts refer to the authors' evaluations of the literature. The adverbials represent their positions with respect to the existing information or other scholars' positions. In (145), the use of *entirely* is attenuated by the downtoner *almost* to indicate the evaluation of the semiotic model carried out by unspecified researchers. The form *fully* shows the authors' assumed position in relation to the extant reviewed literature. Although boosters generally have an intensifying function, this is not the case in this instance. The adverbial appears to attenuate the meaning of the string *fail to explain* to partially reduce the critical stance. Additionally, this allows the authors to take advantage of the theoretical part they are interested in to develop their own argumentation. The instance of *highly* in (147) is substantiated by the use of a large number of evidentials in the form of attribution given in the same utterance.

The downtoners used in this stage are *almost, barely, hardly, merely, nearly, only, partially, partly, scarcely,* and *somewhat,* as in the following excerpts:

(148) Furthermore, aside from being spatially bounded, tourist experiences are socially regulated (Edensor, 2001) and 'tourism is a social experience' (Sharpley, 2002: 315). Likewise, the tourist experience is not *merely* shaped by the physical elements that can be processed by the senses (Walls et al., 2011). It is important to note that the stage where tourists perform is shared with other actors who have specific roles, that the tourist co-produces the experience with (Edensor, 2001). (Aquino *et al.* 2018)

(149) Generally speaking, the above review provides evidence that socio-demographic and economic characteristics are important influencing factors on leisure and tourism expenditure. Yet *only* a handful of studies have been conducted to look at households' spending patterns on tourism products and services; and even less work has been done in comparing tourism expenditure patterns over time and across individuals in China. (Lin *et al.* 2015)

The adverb simply co-occurs with the negation particle *not* to adjust the authors' perspective and enhance the subsequent aspect of the discussion, which is introduced by the evaluative *that*-clause *It is important to note* (. . .). The use of the negative operator takes place with the downtoners *merely* and *only* in the corpus. This collocation with these two adverbials represents 30 percent of the total cases. Boosters also sporadically occur with *not*, but this does not seem to be an outstanding collocation. There is a frequency of 1.9 percent of these instances, and the form encountered is *completely*, as in the following case where theoretical subject matter is discussed:

> (150) Scholars of non-representational theories, however, are not *completely* opposed to any form of representation (Dewsbury et al., 2002; Lormier, 2005; Thrift, 2008) (Prince 2018).

The forms of boosters and downtoners in the research design stage are the same as those in the approach stage. The function of the boosters in this section is to enhance the perspective in the description or evaluation of the method (or a part thereof) or of the research tools, as evinced in the following instances with *very* and *greatly* in the medial position:

> (151) This has *very* real implications for social and political relationships in the region. Well-worn paths of racist discourse can thrive on this moral terrain that reenergizes stereotypes of "the native in the way of progress" (Furniss, 1999; Harding, 2005). (Cooke 2017)

> (152) The demographic profiles of the participants from the two sites are *greatly* diversified across a wide range of age and ethnicity, but the majority of them do have higher education background and stable income (Wang and Alasuutari 2017).

Downtoners are used with a similar function in this section, as in (153) with *somewhat*. Nonetheless, they are also featured as a vague language device to avoid precision or to clearly show the authors' evaluation of a particular feature, as in the examples with *slightly* in (154) and (155):

> (153) All eight brochures and websites analyzed in the following section were published by local government agencies in Qiandongnan and were first collected and/or viewed in 2012 and 2013. I selected these *somewhat* arbitrarily from a larger pool of materials I encountered during this time. (Bratt 2018)

> (154) The walking tour attracted a *slightly* younger audience, as the concerts require a 10-year waiting list, but the tour is open to anyone. Offered in the afternoon, it was also an alternative for tourists who had missed the early morning historical tour of the town. (Bolderman and Reijnders 2017)

> (155) In the past 25 years, there has been no significant population growth, with the population dropping *slightly* from 1,643,542 in 1991 to 1,613,393 in

2014. Conversely, in the past few years, the number of tourists has grown steadily (Bruttomesso 2018).

As to the results stage, boosters are used to describe the way in which a research activity process should be realized, as in (156):

(156) The presence of humans in Figure 3 suggests a reading of signs that emphasises the mobility of the image of Iceland. Notions of corporeal travel and co-presence (see Sheller and Urry, 2004; Urry, 2007) suggest that in order to *fully* apprehend the tourist experience, one must be physically present to negotiate tourism places. (Lund *et al.* 2017)

These devices are also used to qualify findings in this stage:

(157) However, we also observe that the seasonal AR coefficient is *extremely* large in *practically* all cases, implying that seasonality matters. Moreover, the results from Dickey et al. (1984) and Hylleberg et al. (1990) seasonal unit root tests *strongly* support the need for seasonal differencing. Thus, in what follows, we take seasonal first differences on the log transformed data as displayed in Figure 3 (total) and Figure 4 (by region). (Payne and Gil-Alana 2018)

The adverbial *extremely* reflects the stance of the authors; most importantly, this device is also used in an epistemic sense, and accurate information is clearly concealed. The form *strongly* is substantiated by the attribution evidentials given previously in the same utterance. This booster comes to justify the authors' research decisions described in the following sentence and introduced by the consequence logical operator *thus*.

The forms of the downtoners in this stage are not like those in the other stages, save for the use of *practically*, quoted in example (157) above. These devices show an evaluation of the results, too, as in the following excerpt with *partially*:

(158) To test the mediating effects in this study, simultaneous regression paths using AMOS 20 were constructed in our conceptual model (as presented in Figure 2). The simultaneous results show the direct effects of quality–net value (b 1/4 0.35, p< 0.001), price–net value (b 1/4 ¤0.32, p< 0.001) and net value–WTB (b 1/4 0.17, p< 0.01). On accounts of the indirect effects of quality–WTB (b 1/4 0.38, p< 0.001) and price–WTB (b 1/4 ¤0.40, p< 0.001), net value has a *partially* mediating effect on the relationships

between quality and WTB and between price and WTB (all path coefficients are statistically significant at the same level of alpha, $p < 0.01$). Therefore, both hypotheses 1a and 1b are *partially* supported. However, the net value does not mediate the relationship between risk and WTB because the direct and indirect effects of risk to net value and WTB (b 1/4 ◻0.10, ns; b 1/4 ◻0.11, ns, respectively) are not significant. Thus, hypothesis 1c is not supported. Additionally, quality also *partially* mediates the relationship between price and net value (at the same level of alpha, $p < 0.001$), given the indirect effect of price–net value (b 1/4 ◻0.32, $p < 0.001$) and direct effects of price–quality and quality–net value (b 1/4 0.24, $p < 0.001$; b 1/4 0.35, $p < 0.001$, respectively). However, the risk does not mediate the quality–net value relationship because the direct effect of risk to the net value (b 1/4 ◻0.10, ns) is not significant. Based on the above results, hypothesis 2 is *partially* supported, but hypothesis 3 is not supported. (Shiu 2018).

As seen with other boosters, the form *partially* is descriptive to some extent, as its use avoids a more precise term to qualify the findings presented in the study. Precision is, however, revealed in the figures given in the text. Notwithstanding, it is difficult to learn what the authors mean with this adverbial in those cases in which even if there is a clear-cut correlation with figures, the use of *partially* in chunks like *hypothesis 2 is partially supported* is problematic. To say that a hypothesis is partially supported might also imply that the hypothesis is not supported, as the first option is not exclusive of the second. For the authors, however, the use of this adverbial may suggest more successful research perspectives and/or prospects.

The discussion stage offers the same booster and downtoner forms as the results stage, including *practically*. Boosters are deployed to underscore aspects that are assumed to be able to justify the findings under discussion, as in (159):

(159) Importantly, scholars have suggested that gender *greatly* influences barriers to and experiences of leisure and nature tourism (Berryman 2015; Cosgriff *et al.*, 2009; Swain, 1995; Wearing, 1991; Wilson and Little, 2005) (Weatherby and Vidon 2018).

Highlighting specific explanatory reasons in the discussion of findings appears to be the primary target of these stance adverbials in (160) and

(161). In this last example, the booster *highly* and the evidential adverb *clearly* co-occur with a patently indexical function.

> (160) These situational impacts *strongly* correspond to the travel motivations of meeting new people and experiencing foreign cultures (Reichenberger 2017).

> (161) Third, the extent to which these security measures are *clearly highly* racialized cannot be ignored. While less well studied in the Canadian context, racial profiling occurs at international borders (Torabian and Mair 2017).

The downtoners in this stage fulfil the objective of qualifying the results drawn from the study performed. They also contribute to presenting suggestions to improve a negative aspect being discussed, as shown in the following example with *merely*:

> (162) On other occasions, some employees completely flouted the norms of customer service by engaging in violence towards unruly customers. This went beyond *merely* removing disruptive individuals. Instead, there was an unstated assumption that violence against misbehaving customers was a legitimate course of action. (Chapman and Light 2017)

In conclusion, *totally* is to be added to the list of boosters found in the corpus of RAs, as in the following instance:

> (163) When travel agencies move to public ownership, they decrease organization risk and thus form competitive advantages. Nevertheless, some travel agencies do not *totally* succeed in realizing their planned IPO (Huang and Chang 2018).

The use of *totally* preceded by the negative particle *not* has a conclusive tone as a consequence of the evidence that has been obtained in the research carried out. The use of the adverbials *fully, strongly,* and *very* pursues different pragmatic motivations in the following excerpts:

> (164) However, Edensor and Holloway (2008: 486) note that a limitation of Lefebvre's rhythm analysis lies in the fact that in its original formulation, it can never *fully* grasp the manner in which non-human, technological, and material rhythms co-evolve with the body as an assemblage to produce an unfolding of space and time. (Rickly 2017b)

(165) Tourism not only plays a significant role in relation to how musical culture is presented to 'others' (Johnson, 2002); it also *strongly* impacts the development of local musical identity (Fitzgerald and Reis 2016).

(166) Overall, there is still a need for more critical research about this area given the amount of travel that students are now encouraged to do. We are aware that this is a *very small* project but even so recommend that there must be a space for participants to reflexively engage with their tourism experiences beyond sharing photos on Facebook and what Harrison (2008: 42) calls 'cursory description'. (Bone and Bone 2018)

While *fully* and *strongly* report on the authors' impressions concerning some concluding claims and have a clear emphasizing target, the form *very* is used to qualify the word *small,* referring to the project the authors describe. Tangentially, a mitigating function is intended to make clear that they understand that their research is subject to additions, corrections, and criticism. The strength of the research is then rested also on its value as a driving motivation for more research to take advantage of the information student travel experiences may provide.

Downtoners in the conclusion section include *almost, barely, hardly, merely, only, partially, partly, slightly,* and *somewhat.* These are deployed (a) to evaluate the research done or findings, (b) to indicate the limitations of the study presented or any part of it, and (c) to show the authors' perception of the conclusions obtained in relation to existing knowledge. The following examples illustrate these aspects:

(167) Tourist bars and restaurants are *almost* certain to remain the primary supporters of music-making on the island, with other continuing support coming from non-governmental organisations (such as Tamar and Projeto Golfinho Rotador) that intentionally integrate with tourism as they pursue their environmental conservation agendas. (Fitzgerald and Reis 2016).

(168) Nonetheless, as this research *only* examined the representation, promotion and construction of the gap year phenomenon and discourse from a provider's perspective, future research is necessary to explore the gap year experience and (perceived) benefits from the perspective of the gapper, as well as their parents. (Hermann *et al.* 2017).

(169) Among the changes that ensued was the growing access of the Portuguese to leisure and mobility within a developing consumer culture, but also the arrival of growing numbers of foreigners, at once coveted and feared. Filmmakers were eager to engage with these changes. Given the complexity of the contexts involved (of which I have *only* offered an outline), it is

hardly surprising that the representation of tourism should have taken on a variety of shapes and meanings. (Sampaio 2017).

The adverbial *almost* in (167) adjusts the meaning of the epistemic adjective *certain*, and the combination *almost certain* is therefore used as a negative politeness strategy to avoid a strong claim. The form *only* in (168) and (169) refers to the authors' own identification of the limits of their studies. In the case of (168), this identification allows the authors to suggest that further research in the field should be carried out. Finally, the authors use the adverbial *hardly* in the evaluative expression *it is hardly surprising that* (...) to show that the results are to be expected in terms of the contextual variables involved. As such, the results belong to an intersubjective domain and can function as a positive politeness strategy to avoid further counter-reactions.

4.5 Passives

Passive voice is another feature normally associated with academic writing (Banks 2017). The use of this form in the generic stages exhibits statistically significant variation, as shown in Tables 16, 17, and 18.

Variable	F	p (p < 0.05)	F crit	Effect size
Passive voice	12.34	0.047	2.23	0.147

Table 16. Passive voice: One-way ANOVA variance analysis measures.

	Introduction	Approach	Research design
Mean	92.34528	108.0500	155.0728
SD	55.36936	41.1076	92.1921
Minimum	0.0000	0.0000	15.267176
Maximum	224.3589	207.1486	500.7153

Table 17. Passives: Mean, standard deviation (SD), minimum, and maximum values in the introduction, approach, and research design sections.

	Results	Discussion	Conclusion
Mean	102.9647	88.8492	87.1557
SD	42.7818	33.9129	50.0019
Minimum	24.7371	0.0000	0.0000
Maximum	193.3564	171.8834	233.5025

Table 18. Passives: Mean, standard deviation (SD), minimum, and maximum values in the results, discussion, and conclusion sections.

The highest means occur in the research design stage. It appears, both in this methodological stage and the rest of the stages in the RA, that the goal is to take advantage of the allegedly impersonal value of this structure to avoid subjectivity in discourse and promote a sense of personal distance (Reilly et al. 2005: 191; cf. Conrad 2018). Below, I offer examples of passive voice sentences randomly taken from the corpus of RAs:

(170) In this sense, as it will be discussed, many promotional images display apparently static men and women (Vanolo and Cattan 2017) (approach).

(171) The findings are presented under the emotive concepts of commensality and spatiality [...] Both concepts are subdivided by the three aspects of sensory, visible and group dynamics (Schänzel and Lynch 2016) (results).

(172) What appears to distinguish such dishonesty in this case is that it was reported as being a direct response to the behaviour of customers (Chapman and Light 2017) (discussion).

As shown in examples (170) to (172), conceptualizers are not explicitly given in any of the passive forms regardless of the stage in which these sentences appear. In (170), the conceptualizers are the authors of the text (as contextually determined). Responsibility is thus implicit and reliant on the authors. In the case of (171), the focus is placed on specific aspects of their research given in the topic sentence position, namely *findings* and *concepts*, in order to sound more objective and, consequently, to gain credibility. This would also explain the higher means found in the research design stage (see Tables 17 and 18), as this stage is concerned with the presentation of data, participants, and research methods. In the case of (172), the conceptualizer is opaque and cannot easily be retrieved from the available context. As it stands, the passive form *it was reported* refers to shared knowledge, which implies shared responsibility. This pursues the avoidance of explicit accountability and has an attenuating effect to avoid imposition.

As shown in Table 18, the passive voice has lower means in the conclusion and discussion stages, in that order. The reason could be that while the passive voice may occasionally indicate attenuation, there are other more common linguistic resources, such as modal verbs, with a clear scope over the proposition that also mitigate its illocutionary force in several ways. I shall describe illustrative samples for each of the generic stages.

The introduction stage shows samples of passive voice constructions, which are clearly connected to the objectives of this section of the RA, *viz.*, to justify the reason for writing the paper, to give the objectives and the hypothesis, and to put forward the framework and the methods for analysis. The following excerpts show cases of passive voice in introductions:

(173) From a theoretical point of view, the analysis *is grounded* in debates on gendered mobilities and tourism, and it aims at questioning the dichotomy between mobility/empowerment and stasis/disempowerment that *has been discussed* in various contributions in the field of gendered mobilities (cf. Cresswell and Uteng, 2008; Johnston, 2001; Pritchard, 2000; Rose, 1993). (Vanolo and Cattan 2017)

(174) The article *is written* from an anthropological perspective and *limited* to selected national, regional, and local Christian pilgrimages *studied* in Costa Rica. Data *are drawn* from relevant literature and my ethnographic and ethnohistorical work in the Republic, *concerned* with patterns and centers of pilgrimage, over nine periods of field research, from 1984 to 2008. (Straub 2016)

These examples combine basic forms of the passive, e.g., *is grounded* and *are drawn*, with perfective (*has been discussed*) and bare forms (*studied, concerned*). This combination with the bare passives clearly reflects an economical way of developing argumentation from information in the rhematic position. This economical target may also apply to paratactic structures to achieve thematic progression, as in *The article is written (...) and limited* in (174). As already pointed out, the use of such lexical items as *grounded, discussed, written, studied,* and *drawn* is stage-oriented. The authors intend to lay the foundations of the research carried out and presented in the work. This programmatic function of the introduction may favour the appearance of passive structures, as key relevant aspects dealt with in the paper can be thematized. Some of these coincide with the general structure of the work: *analysis* (. . .) *grounded* (theoretical framework and methodology), *dichotomy discussed* (methodology and discussion), and *data* (. . .) *drawn* (methodology). Bare passives do not show this usage and reflect rhematic development, as noted above, in the elaboration of information.

The approach stage comments on the topic of research in relation to the existing literature. In this context, the passive voice allows the topicalization of the subject matter, as in (175) and (176):

(175) Carr (2011) argues that little research *has been conducted* on the social significance of holidays involving parents and children (Schänzel and Lynch 2016).

(176) In a similar vein, according to the classification of Hofstede's (1980) cultural differences, China *is considered* to be a collectivistic society in which the opinions of reference or aspiration groups are more important than individual opinions (Chen and Lamberti, 2015). Therefore, social recognizability *is believed* to be a privilege that Chinese consumers perceive as social status or as a sense of cosmopolitanism. (Correia *et al.* 2018).

Active and passive structures alternate provided the syntactic subject remains the topic being developed, as in the following instance:

(177) *Voluntourism* (Callanan and Thomas, 2005; see also, Chen and Chen, 2011; Conran, 2011; Griffiths, 2015; Raymond and Hall, 2008) *is* a kind of travel *characterised* by the tourist's engagement in volunteer activities [. . .] *Voluntourism represents* both an opportunity and a means of 'value adding' [. . .] While *voluntourism has* many potential benefits, a critical analysis [. . .] (Adams and Borland, 2013; Griffiths, 2016) [. . .] because *voluntourism can be conceptualised* as a place where 'good intentions and imperial designs intertwine', and *questions* 'the recurring dilemmas of acting [. . .]' *Voluntourism has been linked* to a search for fulfilment and personal growth [. . .] *voluntourism can* potentially *amplify* escapist desire by offering new idealistic roles and positionalities [. . .] *voluntourism is studied and framed* (particularly by academics) [. . .] *Voluntourism is* commonly *critiqued* when a lack of critical reflection is not built into the experience. When *voluntourism is positioned* as a mere consumable product, rather than as an important social experience [. . .] *Voluntourism*

> *is theorised* here as becoming and we use notions of cartography and de/
> re/territorialisation as a conceptual framework. (Bone and Bone 2018)

Another fundamental use of the passive voice is concerned with the mode of knowing concerning the information presented, as in these examples:

(178)   A social situation is defined as 'the sum of features of the behaviour sys-
        tem, for the duration of a social encounter' (Argyle *et al.*, 1981: 3) – *these
        social encounters in turn are said to* possess nine distinct yet intercon-
        nected features, whose combination structures and determines social situ-
        ations and the social interactions occurring within. (Reichenberger 2017)

(179)   Hence, *it has been argued* that elements of extraordinary-ness and
        everyday-ness coexist in parts of the tourist experience (Hui, 2008; Shani,
        2013) (Bezzola and Lugosi 2018).

(180)   This area of study is not necessarily new but offers innovative and more
        holistic ways of examining *what has long been observed* (see Cresswell,
        2010a, 2012, 2014; Salazar, 2010) (Rickly 2017a).

(181)   *This situation is believed to* be the reason why the bishop in Mostar has
        condemned the events in Medjugorje (Belaj, 2012) (Kawashima 2016).

(182)   *The IPO process is known* for its asymmetric information among investors
        (Miloud, 2016) (Huang and Chang 2018).

The instances of the passive voice used to suggest the mode or source of knowledge in the approach stage in (178) to (182) are always intersubjective with no clearly identified conceptualizers. In fact, the form *we* is not used in any of these passive structures, and opaque conceptualizers are shown instead, as is evident in all of the instances given. From a pragmatic point of view, these structures function as politeness strategies to protect the authors' public image by attributing the information to the sphere of shared knowledge. The sense of evaluation, however, is evident.

The research design stage shows cases of the passive voice to describe actions related to methodological procedures and the management of data. This is the case in (183) and (184):

(183)   The usual demographic variables, such as age, gender, and marital status,
        were included in the survey so that general explanatory variables could be
        identified and to allow comparisons with other studies. The variables of

travel characteristics *were selected* with reference to other relevant studies (Chen & Chen, 2010; Poria *et al.*, 2004). (Lee 2015)

(184) Data collection proceeded in two stages. In the first stage, a draft questionnaire *was developed* and *tested* to establish the validity of the scales. In the second stage, a formal survey *was* then *conducted*. The empirical data *were collected* from consumers at different spa hotels in four major spa regions of Taiwan during July–August 2013. Customers *were contacted* randomly for the survey during their stay at the hotels. Two hundred questionnaires were issued and successfully *collected* for factor analysis and the subsequent structural equation modelling; this number exceeded a critical sample size of 200 (Garver and Mentzer, 1999). (Shiu 2018)

The reason for the various instances of the passive voice in these two samples is to thematize key aspects of the methodological section: *demographic variables were included, variables of travel characteristics were selected, a questionnaire was developed and tested, a formal survey was conducted*, etc. From a quantitative point of view, out of the 882 passive structures identified in the research design stage, 366 are repeated bundles with lexical verbs often associated with the domain of empirical research.

Examples of the most frequent bundles used in the corpus are *X BE used* (44 cases), *X BE conducted* (33 cases), *X BE made* (16 cases), *X BE formed* (15 cases), *X BE selected* (15 cases), *X BE asked* (14 cases), *X BE based* (14 cases), *X BE considered* (12 cases), *X BE taken* (12 cases), *X BE identified* (11 cases), *X BE viewed* (11 cases), *X BE collected* (nine cases), and *X BE employed* (nine cases), where *X* stands for the syntactic subject.

Other lexical verbs used in this template include *adopted, associated, chosen, considered, found, included, carried, coded, described, designed, excluded, influenced, interviewed, involved, measured, performed, applied, built, located, organized, recorded, written, administered, informed, obtained, undertaken, completed, expressed, extracted, followed, gathered, grounded, implemented, marked, oriented, presented, reported, required, seen, situated, acknowledged, grouped, put, raised, referred,* and *transcribed* among others. All of them, as said before, are strongly connected to actions carried out in operational research practices, and the passive allows authors to place prominence on these aspects.

In the results stage, the instances of structures in the passive voice show lexical verbs related to attestation, exemplification, and evidence among others, as in (185) and (186):

(185) The results of the measurement model confirmed convergent validity and measurement model fit, and the CFA results *are presented* in Table 2. In particular, the high reliability among all constructs *was evinced* by Cronbach's *a* values ranging from 0.83 to 0.98. (Park *et al.* 2016)

(186) The desire for social status *is reflected* in 'Shopping in HK is a symbol of success and prestige' (5.4), in 'Shopping in HK is a social status symbol for me' (5.4) and in 'Shopping in HK means wealth' (5.2). (Correia *et al.* 2018)

The forms found in (185) and (186) fit the description made in the above lines: *are presented, was evinced,* and *is reflected.* In the corpus, the template *X BE presented* is the most common, followed by the templates *X BE used, X BE represented,* and *X BE made,* all describing aspects concerning the evaluation and illustration of the results under inspection.

The discussion stage shows great variation as to the type of the lexical verbs found in passive voice constructions, as shown in (187), to describe and contextualize aspects of research. The evaluative dimension of these structures is observed in (188).

(187) All tours *were* considerably *shaped* by the focus on rudraksha. Guides make it a point to visit a rudraksha tree before later inviting pilgrims to see the one-faced rudraksha in what *is claimed* to be a government centre or shop. The first stop *made* in Kankhal at Anandamayi's tomb *is* clearly *connected* to the rudraksha tree available there. (Aukland 2016)

(188) Further satisfaction positively influences luxury shopping intentions (0.11, $p < 0.05$), which gives rise to hypothesis 7. This result is as *expected* (Correia *et al.* 2018).

As said, unlike the research design stage, the *X BE + PAST PARTICIPLE* template is characterized by the use of a multitude of lexical verbs to fill the past participle paradigm, including *use, base, report, show, observe, require, perceive, identify, find, shape, promote,* and *criticize* among hundreds of other forms. Interestingly, few repetitive sequences are deployed. The most common ones are *X BE used,* with

20 occurrences, *X BE made,* with 16 occurrences, *X BE based,* with 15 occurrences, *X BE considered,* with 12 occurrences, and *X BE seen* and *X BE found,* with 11 occurrences each. All report on certain aspects concerning either the evaluation of the propositional content or the mode of knowing:

(189) While this pattern *has also been found* applicable to other participants of the larger study who were not travelling as backpackers, the advancement to introduce more private details was a phenomenon nearly exclusively observed in the backpacker segment and confirmed by Riley (1988), who emphasized the unusually quick establishment of friendships. (Reichenberger 2017)

(190) In this sense, one could characterize the Berliner neighbourhoods that techno-tourists frequent (i.e. Kreuzberg, Friedrichshain, Neukölln) as neo-Bohemias; consequently, their patterns of tourism *could also be considered* neo-Bohemian (Garcia 2016).

(191) The pilgrimage *is seen* as a catalyst of self-discovery, a journey of reappraisal and transformation of self and one's place in the world (Nilsson and Tesfahuney 2018).

(192) These negotiations of accommodation space *are learnt* from past experiences highlighting the influence of negative memories on present holiday behaviour and the contested nature of holiday spatiality (Schänzel and Lynch 2016).

In the conclusion stage, the passive forms contain lexical verbs related to actions regarding the analysed data, the evidence collected, and the inferences made among other research-based activities:

(193) In this article, we developed a frame analysis of the two largest pilgrimage sites in Bosnia–Herzegovina. Two central arguments *were stated.* First, it *was argued* that the two pilgrimage sites had different tourist and multiethnic frames. Second, it *was maintained* that the structural frames of the two sites produce different forms of guiding, local interpretations and attract different categories of visitors. (Valenta and Strabac 2016)

(194) Our results regarding learning are inconsistent with those of previous studies: No positive relationship *was observed* between knowledge learning and nostalgia (Lee 2015).

As seen in (193), passive voice statements are used in order to elaborate meaning using the communicative matrices *it was argued that*

(...) and *it was maintained* (...) to develop the information given in the previous statement. With this, the authors present the contributions made in their paper. The example in (194) shows that the knowledge has been obtained through observation, and this may give the impression of factual truth regarding the information presented. In this context, the most common template is *X BE found*, which refers to findings. This template also has an evaluative scope on the proposition, as in *The situational features (...) have been found to be a useful framework to analyse the social interactions of backpackers in New Zealand* (Reichenberger 2017), taken from the corpus.

4.6 Conditionals

One-way ANOVA calculation shows that variance in the use of conditionals is statistically significant among the stages in the RAs examined, as shown in Table 19, with higher means in the discussion and results stages (see Tables 20 and 21).

Variable	F	p (p < 0.05)	F crit	Effect size
Conditionals	4.118	0.0011	2.23	0.0054

Table 19. Conditionals: One-way ANOVA variance analysis measures.

	Introduction	Approach	Research design
Mean	4.6685	5.4516	4.1760
SD	9.9971	6.5505	8.7691
Minimum	0.0000	0.0000	0.0000
Maximum	39.3700	24.3632	38.9408

Table 20. Conditionals: Mean, standard deviation (SD), minimum, and maximum values in the introduction, approach, and research design sections.

	Results	Discussion	Conclusion
Mean	10.1041	10.4920	6.3855
SD	11.9233	8.7612	11.9406
Minimum	0.0000	0.0000	0.0000
Maximum	57.3888	37.3599	63.8977

Table 21. Conditionals: Mean, standard deviation (SD), minimum, and maximum values in the results, discussion, and conclusion sections.

Conditionals in academic writing are seen as interpersonal devices to include "the reader in the on-going argumentation" (Warchał 2010: 149), and they are therefore suitable for indicating the authors' stance. This, in turn, may serve to have a persuading effect in the presentation of meaning to the scientific community. Conditionals are presented in the corpus in the following forms: *if, unless, whether, whenever, as long as*, and *given that*. Two examples are presented below:

(195) If we view on-screen tourism as a dynamic place-making phenomenon, we can see how destinations can capitalise on it (Lundberg *et al.* 2018).

(196) Some involve the development of relationships between fans and musicians that would not occur unless both were contained aboard a ship for the duration of the festival (Cashman 2017).

These two examples contain different conditional clauses introduced by *if, whether*, or *unless*. Following Warchał (2010: 143), these conditionals have an ideational meaning, sharing with readers the data and the circumstances surrounding the data. The function of conditionals is to allow the authors and their readers to reach agreement through explicitly indicating logical reasoning in the form of a premise and its likely conclusion (Warchał 2010; Puente-Castelo and Monaco 2013). In so doing, the authors' involvement and commitment to the proposition are clearly manifested; however, the presence of conditionality may imply a lack of imposition over the readers. The highest value in the discussion stage seems to confirm this, as this part of the RA is devoted to presenting interpretations and inferences of the findings. In the introduction, there are conditionals introduced with *if*:

(197)   At this time, between 30% and 40% of the visitors came from other coun-
        tries than Sweden and because of the large international demand for this
        type of tourism (Eagles, 2014a; Garms *et al.*, 2016), it is of special interest
        to study *if there are differences in income elasticities based on nationality.*
        (Fredman and Wikström 2018)

(198)   Some research studies have examined IPOs on the Taiwan Stock Exchange,
        but they ignore travel agencies (Chen, 2012, 2014; Chen and Chen, 2004,
        2010; Huang, 1999; Linda, 1996; Shen and Wei, 2007). Thus, it is an inter-
        esting topic to further investigate *if travel agencies behave differently
        from other travel-related industries when undergoing an IPO in Taiwan's
        stock markets.* (Huang and Chang 2018)

These conditional structures are used to provide the objectives of the
work. They are examples of open conditionals by which the authors
present their research questions within a logical thread of argument.
In the first case, the authors start with factual statistical data to include
what the aim of their research is. The same is true in (198), but the context in which the conditional is inserted contains examples of attribution to justify the need to undertake the research activities. The logical
operator *thus* reflects this necessity and sense of the obvious, and the
evaluation presented in the matrix, i.e., *it is an interesting topic to*, also
reinforces this same idea.

Open conditional operators can also be used in combination with
other question words to specify the exact scope of the existing research
as a means of justifying one's own research. The use of the *if* and *how*
clauses in (199) pursues this aim. In addition, the extensive list of references contextualizes the conclusion in the form of indirect questions:

(199)   In other words, although a number of studies have, for example, explored
        the (secular) spiritual dimension of tourist engagement in specific (reli-
        gious) journeys, such as the Camino de Santiago (Devereux and Carnegie,
        2006), of the experience of religious sites (Francis *et al.*, 2008; Sharpley
        and Sundaram, 2005), of specific settings, such as the countryside (Jepson
        and Sharpley, 2015) and of related experiences, including holistic (Smith,
        2003) and wellness tourism (Steiner and Reisinger, 2006), few have
        addressed *if and how tourists experience a sense of meaning or spiritual-
        ity through tourism more generally* (see Willson *et al.*, 2013). (Jarratt and
        Sharpley 2017)

Open conditionals are also deployed within the formulation of the specific objectives of the RA:

> (200) Specifically, the study has four objectives as follows: (1) examining the mediating influence of quality on the relationship between price and net value, (2) assessing the mediating influence of risk on the relationship between quality and net value, (3) exploring the mediating influence of the net value of the relationship between its antecedents (i.e. quality, price and risk) and WTB and (4) determining *whether demographic variables have moderating effects on the decision paths.* (Shiu 2018)

In (200), the rhetorical conditional is used with one of the aims of the RA to engage epistemic enquiry. Conditionals are also common in the approach stage of the RA, as exemplified in the following instances:

> (201) The concept is associated with Habermas (2006, 2008), although Van Peursen (1989) spoke of the post-secular era way back in 1989. In 1983, Edward Said had already questioned the death of religion in modernity due to secularization and even alerted us to its return. The proponents of the post-secular thesis have even questioned *whether religion was at the margins of modernity.* (Nilsson and Tesfahuney 2018)

> (202) Notably, the actual video was produced in 2012 and was posted on YouTube on Greek Independence Day (Mack, 2012), signalling for young people a performative Grexit as symbolic liberation from troika oppression outside Greece. It is questionable *if the dance event was a nationalist gesture;* in all likelihood, it was escapist creativity mocking at dead heritage custom— an alternative tourist modernity (Ateljevic, 2008). (Tzanelli and Korstanje 2016)

The conditionals in (201) and (202) state the authors' assessment of their review of the existing literature at this stage with a focus on the aspects that require further consideration. The epistemic verb *questioned* in (201) and the epistemic matrix *It is questionable* in (202) signal uncertainty in the elaboration of the justificatory claims.

The below example contains instances of two rhetorical conditionals. The function of these conditionals is to make a strong claim in both cases. Actually, both conditionals are linked paratactically to express two perspectives that, even if they are mutually exclusive, refer to actual truth providing any of the two conditions is met.

(203) Busby and Klug (2001) and Roesch (2009) indicated that people are film
       tourists when they are seeking or searching sites or destinations seen in
       their favorite films or television programs. Tourists may feel dissatisfied *if*
       experiences do not match prior perception and feel satisfied *if the actual*
       experience matches the hyper-real expectation (Carl *et al.*, 2007). (Zhang
       and Ryan 2018)

The open conditional in (204) evidences the logical reasoning used in
the elaboration of information. The formula *if true* given in the thematic
position captures the information presented in the previous statement
as the necessary condition for the claim in the apodosis. However, the
use of *otherwise* at the beginning of the subsequent sentence refers to
the non-fulfilment of the same condition to achieve a different scenario
for decision-making about selected tourist destinations:

(204) In consequence, there must be a hitherto unexplored social force that
       indirectly leads travelers to particular places where other individuals of
       their network go with great inclination, too. *If true,* this implies that per-
       sonal contacts are strong predictors for travel behavior. Otherwise, chance
       meetings would primarily be a product of the attraction of world-famous
       destinations. (Beritelli and Reinhold 2018)

The excerpt in (205) shows the potential of open conditionals to develop
argumentation leading to a claim. The presence of *thus* reinforces the
idea of logical thinking. Overall, making evident the reasoning process
leading to a claim may function as a negative politeness strategy to
avoid imposition on the authors' personal consideration of the existing
literature.

(205) *If the elasticity is positive,* it is a "normal" good (or service) and *if it larger*
       than one, it is also a luxury good. Thus, *if the relative increase in con-*
       sumption is greater than the relative increase in income, it is a luxury
       good. *If the elasticity is negative,* it is an inferior good; a relative increase
       of income leads to a relative decrease in demand. However, in the cases
       when a substantial part of the sample consists of individuals without any
       spending, it is common to use an estimator that uses all information in
       the sample, that is, also those with zero expenditure. (Fredman and Wik-
       ström 2018)

The following are instances of hypothetical conditions:

(206) The number of people engaging each year in this type of activity appears to be rising *if I were to judge solely on the variety of the offers*. It is almost impossible, however, to evaluate the number of visitors because of the total lack of official statistics from the Church or the tourism authorities. (Banica 2016)

(207) Religious faith is the spiritual dimension of Dasein. Religiosity is no longer bound to authority, rather it has become a matter of individual choice. Traditional religions lose authority as their canon is challenged by individuals and religious pluralisms. Religious faith is no longer a given, *if it was ever*. Thus conceived, unlike traditional faith, post-secular religiosity denotes non-hierarchical, self-oriented forms of religiosity (Taylor, 1989; Rowe, 1999; Sigurdsson, 2009). (Nilsson and Tesfahuney 2018)

The hypothetical structures of *If I had to judge* and *if it was ever* help to construct and negotiate meaning. This is achieved by means of perspective signalling, whereby the authors show hesitancy concerning certain aspects of the literature.

In the research design stage, rhetorical conditionals are used to make claims concerning aspects related to the methodology, as in (208) and (209).

(208) Prospective interviewees were told that they would participate in a brief interview about their past travel experiences. For first round interviewees, availability and willingness to participate were the main sampling criteria. We had no prior knowledge *whether the respondents had ever experienced any chance meeting*. (Beritelli and Reinhold 2018)

(209) A narrative interview is appropriate *if one wants to draw out deeply meaningful and personal stories and experiences*. Narratives offer productive ways of capturing the motivations, reasons and investments individuals have in relation to events and place that are of significance to their lives. (Nilsson and Tesfahuney 2018)

The below instance also shows the author's stance by means of a hypothetical condition:

(210) My identity as a tourism studies researcher and tourist is intentionally blurry in this article. Although I explained to the tour company owners and guides that I research tourism at a university, I also insisted on paying my own way as a tourist without any special "favors" attached to my participation. I felt that *if I were to be offered and accept complimentary invitations to join these tours,* I would be seen as favoring certain companies

over others or taking advantage of my positionality as a tourism researcher
to "game" the system. (Gillen 2016)

The conditional structure in (210) reveals a highly unlikely scenario
since it would jeopardize the integrity of the work, which would not be
acceptable in the author's community.

Conditionals in the results stage are used to display statistical
information in logical sequences, as in (211) and (212). With this type
of structure, authors can present information in an analytical way to
support their arguments.

(211) f Equals 1 *if access to cabin or permanent living in the Fuluffall region, 0
       otherwise* (Fredman and Wiksström 2018).

(212) Visitors falling into these two markets segments would face an increase of
       16% rather than just 10% for the average case *if the proposed 10-year visa
       replaces existing visa fees.* As such, the reduction in visitor nights could
       well be at 33.6% rather than 21% reduction as in the above case study.
       (Pham *et al.* 2018)

In the discussion stage, open conditionals are established to introduce
supporting and explanatory evidence according to logical reasoning,
as in (213):

(213) Place branding strategies are linked to reputation: *If a place has a good
       reputation,* it might attract investors and tourists (Jiménez-Esquinas and
       Sánchez-Carretero 2018).

With the following open conditional, the reader is informed through the
classic logical formula. The conditional highlighted here represents an
explicative reformulation of the previous quotative.

(214) For politicians and craftswomen, bobbin lace production '[this] is not
       only about culture, this is also about an economic activity. And if it's not
       helping the economy then there is no point in protecting it' (10 February
       2014, audio recording AU042, employee of the municipality working to
       promote bobbin lace) (Jiménez-Esquinas, 2016). They cannot understand
       the transformation of bobbin lace making into a cultural curiosity for tour-
       ists *if it's not economically profitable.* (Jiménez-Esquinas and Sánchez-
       Carretero 2018)

The hypothetical conditional in (215) helps develop the argument by sharing with readers how the conclusion was reached. The information given is assumed to have an opposite effect, hence the use of *alternatively*. This has a clear descriptive power.

> (215) By immediately fore-grounding their symbolic capital, these displays connote the prestige of both the tour and the guide set out in publicity discourses. Alternatively, *if read in terms of cultural and symbolic power,* the guide's self-presentation recalls strategies identified in previous studies of official media tourism as their clothing constructs an '[i]nvisible, implied boundary' (Couldry, 2000: 107) between themselves and tour participants. (Garner 2017)

In the conclusion section, hypothetical conditionals are used to indicate advice for improving the economy in relation to the tourism industry. As noted previously in chapter 3, this is a function that is often associated with the conclusion in these articles.

> (216) Visas are an effective tool for the governments to ensure secure borders for their countries. However, *if too protective or too revenue-oriented,* visas could be detrimental to the host economy due to potential losses of tourism revenue and foreign direct investment opportunity from overseas countries. (Pham *et al.* 2018)

In (217), the open conditional is deployed to introduce cognitive reasoning, i.e., *If we assume*. The structure relies on verbs often associated with argumentative processes, as they establish logical thinking.

> (217) Due to scarcity of research, we do not really know how tourists – a consumer group mostly comprising individuals living in affluent societies – used to share their travel experiences before the Internet and the arrival of SNSs. *If we assume that online chatting with friends is not entirely unlike people sharing their travel experiences around a kitchen table,* we could argue that SNSs just give us easy access to how it is done in practice. (Wang and Alasuutari 2017)

The use of the following open conditional allows the author's final idea to be presented in contrast to an earlier idea, which is offered as quotative material. The use of the conditional in this case shows the reasoning path, which helps to mitigate possible imposition.

(218) At its most basic, this article shows that if *"place matters in existential authenticity" (Rickly-Boyd, 2013: 683)*, then it also matters in particular places and at scales like the urban (Gillen 2016).

Rhetorical conditionals in this section may be used to recall the objective of the paper. This is the case of the two instances with *whether*. This allows the presentation of the concluding remark given at the end of this excerpt.

(219) This research sought out to prove *whether these new disruptors in the form of low-cost long-haul carriers could replace Charter airlines in the long-haul markets* [. . .] The research aimed to extract *whether low-cost long-haul airlines would replace Charters operating to distant holiday destinations from a passenger perspective.* The results were encouraging as Charters airlines were perceived as a valued entity. (Martín Rodríguez and O'Connell 2018)

Material for future research is introduced by means of rhetorical conditionals in conclusions, as exemplified in (220) and (221):

(220) In fact, it remains to be seen *whether "dark tourism" will hold as an analytical category* (Schäfer 2016).

(221) To continue with this line of research, we suggest to study *whether the presence of foreign direct investment in the hotel industry gives rise to an increase in the competitiveness* of the destination of the investment, due to knowledge transfer. (Mendieta Peñalver *et al.* 2018)

In these examples, the authors present future research prospects by means of *whether* conditionals. These could be categorized as positive politeness strategies to anticipate aspects that may be criticized for not having been considered in the RA that is presented.

4.7 That-*complement clauses*

The use of *that*-complement clauses is statistically significant among the stages in the RAs examined, as shown in Table 22.

Variable	F	p (p < 0.05)	F crit	Effect size
Conditionals	8.2	0.0002	2.23	0.102

Table 22. That-clauses: One-way ANOVA variance analysis measures.

That-complement clauses in my corpus are a means of demonstrating authorial perspectives and estimations concerning the description and evaluation of findings in academic research (Mauranen and Bondi 2003; Stotesbury 2003; Charles 2007; Jalilifar et al. 2018). In line with their functions, the results, discussion, approach, and conclusion stages show higher mean measures than the other stages in the RA, as shown in Tables 23 and 24.

	Introduction	Approach	Research design
Mean	43.1802	65.1697	42.7584
SD	32.0389	33.5147	32.4040
Minimum	0.0000	0.0000	0.0000
Maximum	124.2236	144.6416	148.4781

Table 23. *That*-complement clauses: Mean, standard deviation (SD), minimum, and maximum values.

	Results	Discussion	Conclusion
Mean	69.9247	68.7939	63.5416
SD	29.5439	35.3153	38.0623
Minimum	21.4592	0.0000	0.0000
Maximum	157.6044	166.4145	172.9106

Table 24. *That*-complement clauses: Mean, standard deviation (SD), minimum, and maximum values.

Some examples of *that*-complement clauses are presented below:

(222) *The fact that* such sites are promoted by travel agencies, guidebooks and public tourism authorities and that Hindus today might visit these sites on package tours with more than just religious purposes in mind need not lead us *to conclude that* a more than 2000-year-old tradition has collapsed into tourism. (Aukland 2016)

(223) *I had the impression that* this type of trip would be the ideal terrain to study the existing relationships between tourism and pilgrimage (Banica 2016).

(224) *It may seem unlikely that* a visit to Morecambe, with its image problems and other challenges, may be thought of as a 'spiritual' experience (Jarratt and Sharpley 2017).

(225) *This means that* the intention to shop for luxuries may increase in regard to tourists who are highly motivated by status (Correia *et al.* 2018).

(226) *It has also been reported that* the validity and the reliability of the Delphi technique do not significantly improve with more than 30 participants (Huang and Chang 2018).

(227) Therefore, *it can be concluded that* Charter airlines operating to long-haul holiday markets will not be replaced by long-haul low-cost carriers (Martín Rodríguez and O'Connell 2018).

The evaluative dimension of the *that*-complement clauses in (222)–(227) is found in their matrices (Hyland and Tse 2005b, 2005a; Hyland and Jiang 2018; Kim and Crosthwaite 2019), which may present a complete sentence introduced (a) by an abstract noun (e.g., *fact* and *impression* in [222] and [223], respectively), (b) by an adjective (e.g., *unlikely* in [224]), or (c) by a communicative verb (e.g. *mean* in [225] and *report* in [226]) or cognitive verb (e.g., *conclude* in [227]). The sources of evaluation (Hyland and Jiang 2018: 140) in all of the examples quoted above are either inanimate, as in the abstract entities *fact* and *impression* in (222) and (223), or concealed, as in the remainder of the instances.

While the abstract entity *fact* refers to trustworthiness, the term *impression* seems to signal the opposite, as an impression typically refers to a lack of evidence. In this case, the information given in (223) and framed by the subjective matrix *I had the impression that* lacks epistemic validity. The intention might be to attenuate the propositional content introduced by the *that*-clause. The opaque conceptualizers in examples (224)–(227) may indicate a conscious avoidance of responsibility concerning the truth of the propositional content as well as an avoidance of imposition of their claims on their readers. In the case of the matrix in (227), i.e., *it can be concluded that*, the conceptualizers are the authors of the RA, and responsibility is implicitly attributed.

As for the evaluation performed, the matrices reveal different epistemic statuses of the information being framed and are also indexical of the authors' positions regarding their texts. In the case of (222) and (225), the degree of factuality is openly and unambiguously manifested. In (223) and (224), the matrices specify different degrees of probability concerning the truth of the information presented. These two devices may be used to avoid the authors' imposition of their claims upon their readers. Finally, the communicative and cognitive matrices in (226) and (227), respectively, are used intersubjectively to indicate shared knowledge and, consequently, shared responsibility. All of this works to reinforce the idea that *that*-complement clauses may be used as negative politeness strategies in the RAs studied to avoid the imposition of the authors' point of view in line with some of the other structures revealed here.

The introduction stage shows a preference for *that*-complement clauses with communicative matrices if cognitive and experiential matrices are also found. Communicative matrices include the forms *argue, propose, tell, point out,* and *show* among others, as shown in (228) and (229).

> (228) Music tourism often involves live music events, and *it has been argued that* these events like festivals and concerts make music tourism economically viable (Roberts, 2014) and account for its special feel or appeal – music tourism in this sense is understood as 'travel to hear music played'. (Lashua *et al.* 2014)

> (229) Moreover, *Cunado et al. (2008a) show that* tourist arrivals to the United States exhibit mean reverting behavior (Payne and Gil-Alana 2018).

These instances provide arguments to justify the aims of the authors' research. In (228), an opaque conceptualizer is used to mark intersubjective meaning, while attribution to a third party is displayed in (229). Both strategies contribute to grounding the authors' positions and to showing their participation in the elaboration of the information. In this sense, these epistemic devices show a lower degree of engagement. Samples (230) and (231) include, however, the (inter)subjective devices *I argue* and *We propose,* and both clearly reveal the authors' involvement and full commitment to the information presented in the subordinate clauses.

(230) *I argue that* scale—a geographic idea that can broadly be defined as representations and practices of reality that frame the world (see Herod, 2010)—is pivotal to understandings of existential authenticity because it allows for a more precise means of understanding how host–guest encounters unfold in and shape particular spaces (e.g. Keese, 2011; Silverman, 2012). (Gillen 2016)

(231) *We propose that* experiential corridors and shared heuristics allow getting closer to understanding the social contingencies of on-site movement patterns that drive travelers to certain places, thus enabling chance meetings (Beritelli and Reinhold 2018).

Cognitive matrices in the introduction are also used (inter)subjectively, as disclosed in the below examples.

(232) *It is generally thought that* the nature of a memorial and the meaning attributed to it by a tourist determines, at least in part, the behaviour that is socially appropriate (Mayo, 1988) (Winter 2015).

(233) Nevertheless, *we believe that* it is possible to identify and systematically analyse variations in the forms of guiding, the experiences of journey to the pilgrimage sites and interpretations of the two sites (Valenta and Strabac 2016).

The cognitive matrix combined with the adverbial *generally* in (232) shows shared knowledge that is hereby used to justify the author's claim, and this meaning is reinforced by the attribution to a third party given parenthetically in the final position.

There are some evaluative *that*-clauses introduced by nouns and adjectives, which may include terms such as *fact, argument, study, premise, position, opportunity, view,* and *(un)likely.* The function of these epistemic devices consists of presenting the authors' methodological positions, as in (234) and (235), with a justificatory and/or programmatic scope.

(234) *My hypothesis that* tourism is used as symbol for a "successful city" is based on the premise that tourism has been used to mask urban crises and thereby create a counter-image that serves to envision a better future for a city (see Roche, 1992; Fainstein and Judd, 1999; Greenberg, 2008). (Tegtmeyer 2016)

(235) According to these studies, *it is more likely that* religious tourism actually increases the attractiveness of sacred places. The case study presented

in this article shows that organization and guidance to/at a pilgrimage site provided by a travel agency can revitalize religious practice. (Kawashima 2016)

The approach stage shows diverse forms of evaluative *that*-clauses with a preference for nouns and adjectives in the evaluation source matrix, with a focus on evidentiality and factuality, e.g., *fact, knowledge, report, clear, evident,* and *obvious,* as in the below examples.

(236) *The fact that* they receive and process this guiding in the negotiation of space and experience of place is a major theme of this article (Banica 2016).

(237) *Thus, it is evident that* the Arctic is a "place in play" (Sheller and Urry, 2004) in which heterogeneous inhabitants, humans and non-humans, are in a constant process of being located and relocated, composed and recomposed in landscapes of constant flow and flux depending on how they are being performed and by whom. (Lund *et al.* 2018)

The reason for the use of evidential lexical nouns and adjectives is related to the functional target of the approach stage. Since this stage is devoted to contextualization of the research presented, a review of the literature in the field is necessary. Therefore, the use of these lexical items is motivated by the fact that they qualify that the ideas have been researched already. The same is true for the preference of communicative evidential matrices deployed, as they allow authors to capture what others say about the topic under review. Common verbs in these matrices are *assert, claim, explain, indicate, note, report, determine,* and *say* among others. Communicative evidentials are useful for presenting third-party ideas that can support the authors' perspective. In doing so, epistemic attribution is revealed to characterize the authors' participation in the elaboration of the information, as evinced in (238), (239), and (240), which show cases of a third-party attribution, intersubjective claims with an opaque conceptualizer, and intersubjective positioning with first-person plural marking, respectively.

(238) *Urry (2003) argues that* co-presence and with it chance meetings are no product of happenstance (Beritelli and Reinhold 2018).

(239) Hence, *it has been argued that* elements of extraordinary-ness and everyday-ness coexist in parts of the tourist experience (Hui, 2008; Shani, 2013) (Bezzola and Lugosi 2018).

(240) In fact, *we note that* the shift from pop and even 'kitsch' imaginaries of happiness (the beach, party and film-induced tourisms for which Greece became known after the 1960s) to dark and heritage-inspired ones, might prompt a controversial status upgrade in global tourist markets. (Tzanelli and Korstanje 2016)

Cognitive matrices may include *assume, seem (that), understand, know,* and *think,* as seen in (241) and (242), which show an indeterminate conceptualizer and an intersubjective conceptualizer, respectively. Both cases represent examples of inferential communication.

(241) *One could assume that* just by being at a place and simply living a whole life there produce a slight chance of meeting the whole world (Beritelli and Reinhold 2018).

(242) Interestingly, *it seems that* the agency, in collaboration with the Catholic priest, is actually able to offer a more relaxed pilgrimage than the one offered by a lay spiritual leader (Kawashima 2015).

In the case of the research design stage, the evaluation of the entities involves nouns, such as *assumption, fact, evidence, probability,* and *perception,* and adjectives, such as *apparent, safe, clear, interesting, important,* and *possible,* as in the following instances:

(243) Given its scale and *the fact that* the Czechoslovak Trade Union was responsible for the construction expenses, this project required an intense level of collaboration (Tchoukarine 2016).

(244) However, *it became clear that* the intention of capturing lived experiences without imposing cultural meanings is a challenging task, when oneself has been raised and disciplined in a European tradition of appreciating mountain landscape (Gyimóthy 2018).

(245) Although TripAdvisor supports uploads of reviews in the Thai language, there were no reviews in Thai and there is no TripAdvisor site in Thai. Thus, *it is safe to conclude that* the data here pre-dominantly represent the perspectives of English-speaking Western tourists, a limitation of the study. (Walter 2016)

Communicative matrices are also common in this stage to describe the research methodology followed by the analysis of the findings or even to raise concerns pertaining to the method used, as in the below instances.

(246) *I argue that* this theoretical framework can approach complex phenomena, such as tourism, in which different socio-spatial dimensions coexist and co-function, either concordantly or in conflict (Briassoulis, 2017) (Brut-tomesso 2018).

(247) *We can also say that* the analysis focuses on a moral aspect of language use regarding authenticity (Wang and Alasuutari 2017).

Evaluation sources in the *that*-clauses in the results section include nouns and adjectives related to the expression of evidentiality and factuality, e.g., *fact, evidence, report, clear,* and *apparent,* as in (248). Epistemic probability items include *possible, probable,* and *unlikely,* as in (249). In agreement with the function of this stage, i.e., the presentation of results, figures, and facts, words related to factuality are more common than those related to probability.

(248) *Our argument that* Iceland is a gateway destination is strengthened by the notion that time–space compression is couched within the MyStopover campaign (Lund *et al.* 2017).

(249) While the government could justify the sunk cost for the first 3 years for the long-term revenue objective, *it is unlikely that* tourism businesses can sustain 3 years with losses for the government to reap the tax benefit, as nearly 95% of tourism businesses are in the form of self-employed, micro and small sizes (TRA, 2016c). (Pham *et al.* 2018)

The same is true of lexical verbs in the evaluation source of *that*-clauses, with examples including *indicate, suggest, argue, show,* and *reveal.* All of these matrices have an indexical function and present the data obtained in the analysis the authors have performed, as in the below samples.

(250) The linguistic ritual manifested as a highly formalized practice evidenced by a limited range of words, phrases and themes as Bell (2009) has described. Of the 2166 VBE, 62% contained at least one of five phrases, which accounted for 52.7% (n = 1498), of the 2842 phrases analyzed here. These were: RIP, Rest in Peace, Never forgotten, Lest we forget and

Thanks/Thank you. *The last column in* Table 2 *shows that* with the exception of Canada and New Zealand, these five phrases accounted for at least 50% of phrases used by all national categories. (Winter 2015)

(251) *The results revealed that* most of the tourists visited South Tainan Railway Station by motorcycle (48%) or by car (33%), and 46 per cent of respondents obtained information about South Tainan Railway Station from their relatives and friends (Lee 2015).

(252) *Participants' narratives indicate that* the spirituality of the journey was constructed by facing the physical challenge of trekking Mount Pinatubo (Aquino *et al.* 2018).

The discussion stage also shows a preference for communicative matrices. The evaluation source includes nouns and adjectives referring to factual truth, e.g., *fact, clear,* and *apparent,* as in (253) and (254). In addition to the presentation of facts, communicative matrices allow the presentation of the authors' ideas, which can be mitigated by the addition of epistemic and dynamic devices, as in (255). In doing so, the aim of the discussion stage of the data obtained unfolds with a combination of expository and argumentative rhetorical strategies, such as the devices I have outlined here.

(253) A classical stereotype analysed in the literature (Buunk et al., 2002), for example, is *the idea that* an old man's physical unattractiveness may be compensated by high social status, education or income (Vanolo and Cattan 2017).

(254) *It is thus clear that* the puffin takes on different materialities in terms of how it appears in the urban landscape as linked with tourism as profitable industry, which urges us to explore further its material composition, or the threads it entangles (Ingold, 2011). (Lund *et al.* 2018)

(255) The shrine undermines parochial sentiments, and *one might claim that* international pilgrims 'bring the world' and development to an area that initially was backward and underdeveloped. Yet, *it can also be argued that,* in its local, Croatian frame of interpretation, the aforementioned transnational and cosmopolitan 'emotional unity with national difference' indicated by Halemba does not include other local, non-Croat populations. (Valenta and Strabac 2016)

Interestingly, attitudinal expressions are also found in this stage in the evaluation source of the *that*-clauses with the use of lexical adjectives, such as *interesting* and *(un)surprising.* These structures clearly mark the authors' stances but can also be felt as a negative impoliteness threatening the addressees' negative face wants, as Culpeper (2011) points out. Their use seems, however, to show the authors' knowledge of the subject matter, and consequently, authority seems to be implied, as in (256), where it is only attenuated by the adverbial *In these conditions* in the left periphery.

> (256) Certainly, the comportment of customers was frequently a source of disappointment and antagonism for employees. In these conditions, *it is unsurprising that* the spirit of the carnival can 'rub off' on employees so that some workers adopted practices which mirrored those of visitors. (Chapman and Light 2017)

In the conclusion stage, *that*-clauses with nouns and adjectives referring to factuality are used to elaborate information from the evidence collected during research, as in the below instances with *fact, indication, evidence,* and *evident.*

> (257) *The fact that* females represented 75% of respondents also presents some issues concerning the relevance of the results to the male population (Salvatierra and Walters 2017).

> (258) The refectory and its management are *an indication that* operators are aware of *the fact that* the authentic "sells" and attracts the tourists just as much as the pilgrim (Banica 2016).

> (259) There is growing *evidence that* Chinese visitors are sensitive to price changes to most of the long-haul destinations and towards Australia in particular (Habibi and Rahim, 2009; Pham *et al.*, 2017; Schiff and Becken, 2011) (Pham *et al.* 2018).

> (260) Even with the negative sociocultural impacts of tourism and the discomfort this at times caused, *it was evident that* interviewees had an exceptionally committed attitude towards catering to the needs of international tourists and providing exceptional levels of service. (Nelson and Matthews 2018)

These evaluative *that*-clauses are also useful to clarify the authors' epistemological stances in their research papers, as in the following example:

(261) This study adopted *the view that* Tyne Cot Cemetery facilitates the performance and rehearsal of memory as ritual practice (Bell, 2009; Duncan, 1995) (Winter 2015).

Communicative evidential matrices also show the authors' involvement in the unfolding of new information concerning the subject matter of the RA.

(262) *I argue that* it is a triumphant and celebratory story of survival that deserves greater scholarly attention (Johinke 2018).

(263) *This research shows that* not only the music plays an important role (Bolderman and Reijnders 2017).

(264) Indeed, *the outcomes of this research suggest that* similar experiences may occur among tourists in other resorts and coastal locations, that is, the research points to the possibility of a transferable seaside experience with spirituality at its core, distinguished by perceptions of timelessness and nothingness, rather than something place specific and bound only to a particular resort. (Jarratt and Sharpley 2017)

(265) *It can thus be suggested that* the backpacker subculture existent in these geo-graphically defined areas may be as much a state of mind as a practical and spatial arrangement, and that the proposed benefits of enclaves such as recovery from unfamiliar surroundings, acclimatization and convenience (Howard, 2007) are in need of further examination. (Reichenberger 2017)

(266) While the latter goal might be seen as a natural response to tourists when one benefits economically from their presence, *we have suggested that* it also manifests from a deep rooted sense of duty linked to Japanese cultural identity. (Nelson and Matthews 2018)

Some of the matrices are used subjectively, as in (262), or intersubjectively, as exhibited in (263), (264), (265), and (266). Some of the intersubjective conceptualizers are implicitly construed, as in (263) and (264), and other matrices present explicit conceptualizers, as in (266), with the pronoun *we* referring to the authors. The use of the modal in (265) pursues to attenuate the force of the claim so that authorial imposition is minimized. The use of the logical operator *thus* contributes to mitigating the statement presented. That is not the case, however, of the instance in (264), as the evidential *indeed* strengthens the authors' perspective. In the conclusion, cognitive evaluative *that*-clauses do not seem to occur beyond *think*, as in (267), and *seem*, as in (268).

(267) In line with the work of Collins-Kreiner and Nurit (2000), I *think* that they refused to define themselves as "tourists" because of the influence of their travel mates (pp. 55–67) (Banica 2016).

(268) As such, *it seems that* partnerships across and within theory and practice hold the key to the successful management of on-screen tourism development and its study (Lundberg *et al.* 2018).

The use of subjective *I think* to show perspective is softened by the use of attribution material to avoid imposition of the claim. In the case of the matrices with *seem*, all of them are construed intersubjectively using opaque conceptualizers, as in (268).

4.8 Conclusions

This chapter explored the use of some language devices used to evince authors' stances in RAs in the domain of tourism. The analysis conducted showed that there are certainly differences among stages concerning the use of some stance features. Tense appears to have some connection with a stage's target, and that explains the use of past tense in the approach stage in contrast to the introduction and the conclusion stages, which have a preference for the present tense. The functions of these tenses range from creating a credibility space for the presentation of research to expressing the epistemic validity of the propositional content.

Modal verbs also present with meaningful variation among genre stages with higher frequencies in the discussion and the conclusion stages. The main functions of modals are to develop argumentation, to suggest politeness, and to express advice. The presence of an effective modality, e.g., deontic necessity, plays a key role in inducing the transfer of research to industries or institutions engaged in promoting tourism. As to boosters, they are used to adjust the degree of writers' commitment to epistemic certainty. These devices present variance as to their presence in genre stages. Boosters appear to be more common than downtoners in RAs.

The use of passive voice in RAs displays relevant variation among stages, being more frequent in the research design stage. Passive voice structures have been found to function as an attenuating strategy by placing agents in the focus position. Arguably, the ultimate goal is to avoid full accountability. It seems that passive voice structures may suggest intersubjective meaning. Conditionals also show statistical significance among genre stages and are used to avoid imposition. By evincing the way in which a conclusion is achieved through logical reasoning, writers appear to minimize the illocutionary force of the claims made. Finally, variation concerning *that*-complement clauses is significant among stages. These structures have an evaluating dimension, and it is therefore unsurprising to find higher mean frequencies in the more argumentative stages of the RA. These structures are used, among other purposes, to disclose attribution in order to define such matters as involvement and responsibility.

5 Conclusion

The main purpose of this study was to identify the macrostructure of the RA in the field of tourism along with the language associated with each part of the RA. Although the broader genre has been described in detail in the literature across several disciplines, this has not been the case for tourism RAs, as noted by Lin and Evans (2012). I have argued in this work that the relative youth of the discipline might impact the way in which empirical findings are exchanged within relevant academic papers and that the discipline itself may also impose its own constraints (e.g., the methodological practices followed in the collection of evidence). I have further argued that the impression of a lack of structural uniformity reported in the literature may be a consequence of the narrative style used in the account of the methodological procedures being followed and in the section headings. Despite all of the above, I have been able to determine the generic structure potential of tourism RAs; notably, this potential does not differ substantially from the traditional Swales' IMRD model (1990).

My characterization of the RAs in my corpus comprises a number of generic stages: the introduction stage, which is obligatory, and the subsequent stages, which are optional, although the approach, discussion, and conclusion stages appear often. It should be noted that all papers analysed in this research present, at least, information on the research topic, the object and method of inquiry, the results, and concluding remarks; however, aspects of these may appear as a single stage or may be mentioned within another stage without actually becoming a single stage.

I have noted in this volume that authors' use of headings to label each stage of the RA is often but not always helpful in identifying the RA stages. These headings may be either functional or narrative. While the former type tends to be made up of institutionalized terms or phrases and clearly reflect what the part of the RA in question is designed to convey, the latter type tends to be both more original and less descriptive in terms of the readers' informational expectations. In the absence of clear visual and linguistic cues in headings to justify my division into stages, I found support from frequent language templates for each genre stage that revealed the specific phraseology in each stage.

This study also assessed the variation and function of perspectivizing features in tourism-related RAs per genre stage. Prior to this account, a description of the structure and language of these RAs was done to ensure that I had accurately segmented the texts into stages. My characterization of the RAs in my corpus comprises a number of generic stages, namely the introduction, approach, research design, results, discussion, and conclusion. Headings were sometimes helpful in identifying the stages. In the absence of clear visual and linguistic cues in headings to justify my division into stages, I found support from frequent language templates for each genre stage, which revealed the specific phraseology in each generic stage. This was further assisted by the calculation of the lexical and informational density and syntactic elaboration per stage, as the highest means of these matched to the stages with a patent interpretative function, in which logical reasoning and evaluation may involve the use of subordinating structures.

My analysis of the variation of a set of perspectivizing features in tourism-related RAs revealed that except for downtoners, the use of the features analysed is statistically significant per stage. The study reveals what functions each of these structures fulfil as stance taking devices. The present tense is used to maximize objectivity in cases where facts and findings are being presented. Authors also deploy this tense along with experiential, communicative, and cognitive lexical verbs as a legitimizing device to suggest epistemic validity, both subjectively and intersubjectively. For its part, the past tense appears in narrative passages to imply that careful and reliable methodological procedures were used in the collection of data. Perspective is also gained using modal verbs, whose main pragmatic function is to be a negative politeness device. However, another distinct function of these verbs is to reveal the necessity of taking specific actions to improve tourism market results in light of the evidence put forward by the research conducted. In this context, the authors openly show their commitment to their claims, as confirmed by their own empirical data-based research and their professional expertise. The tourism and hospitality industry is thus nurtured by these suggestions, based as they are on empirical findings, which seek to increase competitiveness and to set out future directions in the sector.

In line with necessity modal verbs, boosters suggest reinforcement of the authors' point of view, and downtoners function in an epistemic sense to suggest different degrees of certainty. In the case of passive voice, its main perspectivizing function in my RAs is to lessen authorial commitment regarding claims. This is achieved either through the use of opaque conceptualizers or by avoiding the presence of the agent to imply shared knowledge and, consequently, shared responsibility. The conditionals and the *that*-complement clauses in my corpus have important indexical functions, namely to show elaboration of meaning (in the case of the former) and evaluation (in the case of the latter). Both devices seem to be used as negative impoliteness strategies in my texts to avoid the imposition of the authors' stance on the readers.

5.1 Limitations of research

This study has certain limitations that may be considered in future research prospects. Even though I tried to account for the generic structure of the tourism-related RAs, I only managed to provide a description of the RA macrostructure in this paper, so an account of the genre segmentation in terms of the microstructure seems in order so that additional rhetorical and language variation may become evident. This could also consider the existing sub-registers in the tourism literature, which may evince some variation concerning the tourism subfield to which the RA belongs. Similarly, the potential differences concerning the language nativeness of the authors should be taken into account to see whether variation applies. This wealth of information may assist with pedagogical considerations regarding the teaching and learning of the aspects involved in the linguistic elaboration of scientific knowledge in this register.

List of figures

List of tables

References

Abdollahzadeh, Esmaeel 2011. Poring over the findings: Interpersonal authorial engagement in applied linguistics papers. *Journal of Pragmatics* 43/1, 288–297.

Ahmed, Shabbir 2015. Rhetorical organization of tourism research article abstracts. *Procedia - Social and Behavioral Sciences* 208, 269–281.

Ai, Haiyang / Lu, Xiaoye 2010. A web-based system for automatic measurement of lexical complexity. *Paper Presented at the 27th Annual Symposium of the Computer-Assisted Language Consortium (CALICO-10)*. Amherst, MA. June 8–12.

Ai, Haiyang / Lu, Xiaoye 2013. A corpus-based comparison of syntactic complexity in NNS and NS university students' writing. In Ana Díaz-Negrillo / Nicolas Ballier / Paul Thompson (eds) *Automatic Treatment and Analysis of Learner Corpus Data*. Amsterdam: John Benjamins, 249–264.

Aijmer, Karin 2016. Revisiting actually in different positions in some national varieties of English. In Francisco Alonso-Almeida / Laura Cruz García / Víctor González Ruiz (eds) *Corpus-based studies on language varieties*. Bern: Peter Lang, 115–143.

Alonso-Almeida, Francisco 2015a. Introduction to stance language. *Research in Corpus Linguistics* 5/3, 1–5.

Alonso-Almeida, Francisco 2015b. On the mitigating function of modality and evidentiality. Evidence from English and Spanish medical research papers. *Intercultural Pragmatics* 12/1, 33–57.

Alonso-Almeida, Francisco 2012. Sentential evidential adverbs and authorial stance in a corpus of English computing articles. *Revista Española de Lingüística Aplicada* 25/1, 15–32.

Alonso-Almeida, Francisco / Álvarez-Gil, Francisco J. 2021a. Impoliteness in women's specialised writing in seventeenth-century English. *Journal of Historical Pragmatics* 22/1, 121–152.

Alonso-Almeida, Francisco / Álvarez-Gil, Francisco J. 2021b. Evaluative that structures in the Corpus of English Life Sciences Texts. In Isabel Moskowich / Inés Lareo / Gonzalo Camiña (eds) *"All families and genera": Exploring the Corpus of English Life Sciences Texts.* Bern: John Benjamins, 227–248.

Alonso-Almeida, Francisco / Álvarez-Gil, Francisco J. 2021c. The discourse markers indeed, in fact, really and actually and their Spanish equivalents in economy. *Revista de Lingüística y Lenguas Aplicadas* 16/1, 11–23.

Alonso-Almeida, Francisco / Álvarez-Gil, Francisco J. 2019. Modal verb categories in CHET. In Isabel Moskowich / Begoña Crespo / Luis Puente-Castelo / Leida Maria Monaco (eds) *Writing History in Late Modern English: Explorations of the Coruña Corpus.* Bern: John Benjamins, 150–165.

Alonso-Almeida, Francisco / Carrió Pastor, María Luisa 2015. Sobre la categorización de *seem* en inglés y su traducción en español: Análisis de un corpus paralelo. *Revista signos* 48/88, 154–173.

Álvarez-Gil, Francisco J. 2021. Authority and deontic modals in Late Modern English. In Isabel Moskowich / Inés Lareo / Gonzalo Camiña (eds) *"All families and genera": Exploring the Corpus of English Life Sciences Texts.* Bern: John Benjamins, 249–264.

Álvarez-Gil, Francisco J. 2020. A disciplinary analysis of fairly in late modern English scientific writing. In María Luisa Carrió Pastor (ed.) *Corpus Analysis in Different Genres.* London: Routledge, 93–107.

Álvarez-Gil, Francisco J. 2018a. Epistemic modals in early Modern English history texts. Analysis of gender variation. *Revista de Lingüística y Lenguas Aplicadas* 13/1, 13–20.

Álvarez-Gil, Francisco J. 2018b. *Adverbs ending in -ly in late Modern English. Evidence from the Coruña Corpus of History English texts.* Valencia: Universidad Politécnica de Valencia.

Álvarez-Gil, Francisco J. 2017. Apparently, fairly and possibly in the Corpus of Modern English History Texts (1700–1900*).* In Francisco Alonso Almeida (ed.) *Stancetaking in Late Modern English Scientific Writing. Evidence from The Coruna Corpus: Essays in Honour of Santiago González y Fernández-Corugedo.* Valencia: Universidad Politécnica de Valencia, 93–105,

Álvarez-Gil, Francisco J. / Bondi, Marina 2021. Metadiscourse devices in academic discourse. *Revista Signos* 54/106, 518–528.

Álvarez-Gil, Francisco J. / Domínguez Morales, María Elena 2021. Modal verbs in academic papers in the field of tourism. *Revista Signos* 54/106, 549–574.

Álvarez-Gil, Francisco J. / Domínguez Morales, María Elena 2018. Modal verbs in the abstract genre in the field of tourism. *Revista de Lenguas para Fines Específicos* 24/2, 27–37.

Álvarez-Gil, Francisco J. / Payet, Karine Marie Muriel / Sánchez Hernández, Ángeles 2020. Analyse de la modalité dans quelques textes touristiques aux Îles Canaries. *Studii și cercetări filologice. Seria Limbi străine aplícate* 19, 9–20.

Banks, David 2017. The extent to which the passive voice is used in the scientific journal article, 1985–2015. *Functional Linguistics* 4/1, 1–17.

Basturkmen, Helen 2012. A genre-based investigation of discussion sections of research articles in dentistry and disciplinary variation. *Journal of English for Academic Purposes* 11/2, 134–144.

Bazerman, Charles 1988. *Shaping written knowledge: The genre and activity of the experimental article in science.* Madison: University of Wisconsin Press.

Bazerman, Charles 1994. Systems of genres and the enactment of social intentions. In Aviva Freedman / Peter Medway (eds) *Genre in the New Rhetoric.* London: Routledge, 79–101.

Beeching, Kate / Detges, Ulrich (eds) 2014. *Discourse functions at the left and right periphery: Crosslinguistic investigations of language use and language change.* Leiden, The Netherlands: Brill.

Benkenkotter, Carol / Bhatia, Vijay Kumar / Gotti, Maurizio (eds) 2012. *Insights into academic genres.* Bern: Peter Lang.

Bhatia, Vijay Kumar 1993. *Analysing genre: Language use in professional settings.* London: Routledge.

Biber, Douglas 1988. *Variation across speech and writing.* Cambridge: Cambridge University Press.

Biber, Douglas / Gray, Bethany 2011. Grammatical change in the noun phrase: The influence of written language use. *English Language & Linguistics* 15/2, 223–250.

Biber, Douglas / Johansson, Stig / Leech, Geoffrey / Conrad, Susan / Finegan, Edward 1999. *Longman Grammar of Spoken and Written English.* London: Longman.

Biber, Douglas / Johansson, Stig / Leech, Geoffrey / Conrad, Susan / Finegan, Edward 2021. Grammar of Spoken and Written English. Amsterdam: John Benjamins.

Bolinger, Dwight 1972. *Degree words.* Berlin: De Gruyter Mouton.

Bongelli, Ramona / Riccioni, Ilaria / Burro, Roberto / Zuczkowski, Andrzej 2019. Writers' uncertainty in scientific and popular biomedical articles. A comparative analysis of the British Medical Journal and Discover Magazine. *PLoS ONE* 14/9, 1–26.

Boye, Kasper / Harder, Peter 2009. Evidentiality: Linguistic categories and grammaticalization. *Functions of Language* 16/1, 9–43.

Brett, Paul 1994. A genre analysis of the results section of sociology articles. *English for Specific Purposes* 13/1, 47–59.

Brezina, Vaclav 2018. *Statistics in Corpus Linguistics*. Cambridge: Cambridge University Press.

Brezina, Vaclav / Weill-Tessier, Pierre / McEnery, Anthony 2020. *#LancsBox v. 5.x. [software]*. Retrieved from http://corpora.lancs. ac.uk/lancsbox (Accessed 10 July 2021)

Brezina, Vaclav / McEnery, Anthony / Wattam, Stephen 2015. Collocations in context: A new perspective on collocation networks. *International Journal of Corpus Linguistics* 20/2, 139–173.

Brown, Penelope 2011. Color Me Bitter: Crossmodal Compounding in Tzeltal Perception Words. *The Senses and Society* 6/1, 106–116.

Brown, Penelope / Levinson, Stephen 1987. *Politeness: Some universals in language usage*. Cambridge: Cambridge University Press.

Brown, Keith / Miller, Jim 2016. *A Critical Account of English Syntax. Grammar, Meaning, Text*. Edinburgh: Edinburgh University Press.

Cacchiani, Silvia 2018. If-Conditionals in Economics Research Articles: From Keywords to Language Teaching/Learning in the L2 Writing-for-Publication Class? *Corpus Pragmatics* 2/1, 1–26.

Caffi, Claudia 1999. On mitigation. *Journal of Pragmatics* 31, 881–909.

Caffi, Claudia 2007. *Mitigation*. London: Elsevier.

Carrió Pastor, María Luisa 2017. Mitigation of claims in medical research papers: A comparative study of English and Spanish-language writers. *Communication & Medicine* 13/3, 249–261.

Carrió Pastor, María Luisa 2012. A contrastive analysis of epistemic modality in scientific English. *Revista de Lenguas para Fines Específicos* 18, 45–70.

Chafe, Wallace 1986. Evidentiality in English conversation and academic writing. In Wallace Chafe / Johanna Nichols (eds) *Evidentiality: The linguistic coding of epistemology*. Norwood: Ablex, 261–72.

Channell, Joanna 1994. *Vague language.* Oxford: Oxford University Press.

Charles, Maggie 2007. Argument or evidence? Disciplinary variation in the use of the Noun that pattern in stance construction. *English for Specific Purposes* 16/2, 203–218.

Chen, Mingfang 2009. Tense of Reporting in Dissertation Literature Reviews. *Journal of Cambridge Studies* 4 /2, 139–150.

Conrad, Susan 2018. The use of passives and impersonal style in civil engineering writing. *Journal of Business and Technical Communication* 32/1, 38–76.

Cornillie, Bert 2009. Evidentiality and epistemic modality. On the close relationship between two different categories. *Functions of Language* 16/1, 44–62.

Cornillie, Bert / Delbecque, Nicole 2008. Speaker commitment: Back to the speaker. Evidence from Spanish alternations. *Belgian Journal of Linguistics* 22, 37–62.

Culpeper, Jonathan 2011. *Impoliteness: Using Language to Cause Offence.* Cambridge: Cambridge University Press.

de Branbater, Philippe / Dendale Patrick 2008. Commitment: The term and the notions. *Belgian Journal of English Linguistics* 22/1, 1–14.

Denison, David 1993. *English Historical Syntax: Verbal Constructions.* London: Longman.

Depraetere, Ilse / Langford, Chad 2020. *Advanced English Grammar. A Linguistic Approach.* London: Bloomsbury.

Díaz-Redondo, María 2021. Un análisis crítico del uso de pasivas y cláusulas de relativo en los artículos de investigación en "Freshwater Ecology." *Revista de Lenguas para Fines Específicos* 27/1, 140–156.

Diewald, Gabriele / Kresic, Marijana / Smirnova, Elena 2009. The grammaticalization channels of evidentials and modal particles in German: integration in textual structures as a common feature. In Maj-Britt Mosegaard Hansen / Jacqueline Visconti (eds) *Current trends in diachronic semantics and pragmatics.* Leiden, The Netherlands: Brill, 189–209.

Ding, Daniel 2002. The passive voice and social values in science. *Journal of Technical Writing and Communication* 32/2, 137–154.

Dixon, Robert M. W. 2005. *A semantic approach to English grammar.* Oxford: Oxford University Press.

Dzung Pho, Phuong 2013. *Authorial stance in research articles: examples from applied linguistics and educational technology.* New York: Palgrave Macmillan.

Eggins, Suzanne 1994. *An introduction to systemic functional linguistics.* London: Pinter Publishers.

Eisenberg, Peter 2013. *Grundriss der deutschen Grammatik. Der Satz* (4th ed.) Stuttgart: Metzler.

Fang, Zhihui 2005. Scientific literacy: A systemic functional linguistics perspective. *Science Education* 89/2, 335–347.

Ferdinandus, Adelce S. 2016. English Tense Use in Indonesian Journal Articles. *KnE Social Sciences* 1/1.

Fest, Jennifer 2015. Corpora in the Social Sciences – How corpus-based approaches can support qualitative interview analyses. Revista *de* Lenguas *p*ara Fines Específicos 21/2, 48–69.

Fontaine, Lise 2013. *Analysing English grammar: A systemic functional introduction.* Cambridge: Cambridge University Press.

Gil-Salom, Luz / Soler-Monreal, Carmen 2009. Interacting with the reader: Politeness strategies in engineering research article discussions. *International Journal of English Studies* 9, 175–189.

Godnič Vičič, Šarolta 2015. Variation and Change in the Grammatical Marking of Stance: The Case of that-Complement Clauses in Research Articles. *ELOPE: English Language Overseas Perspectives and Enquiries* 12/2, 9–28.

Gotti, Maurizio 2021. Scientific communication in English as a second language. In Michael J. Zerbe / Gabriel Cutrufello / Stefania Maci (eds) *The Routledge Handbook of Scientific Communication.* London: Routledge, 145–155.

Gotti, Maurizio 2017. English as a lingua franca in the academic world: Trends and dilemmas. *Lingue e Linguaggi* 24, 47–72.

Gotti Maurizio (ed.) 2012. *Academic Identity Traits: A Corpus-based Investigation.* Bern: Peter Lang.

Gotti, Maurizio (ed.) 2009. *Commonality and individuality in academic discourse.* Bern: Peter Lang.

Gotti, Maurizio 2003. *Specialized discourse: Linguistic features and changing conventions.* Bern: Peter Lang.

Eisenberg, Peter 2013. Grundriss der deutschen Grammatik. Der Satz (4th ed.) Stuttgart: Metzler.

Halliday, Michael Alexander Kirkwood 1978. *Language as Social Semiotic: The Social Interpretation of Language and Meaning.* Baltimore, MD: University Park Press.

Halliday, Michael Alexander Kirkwood / Martin, James 1993. *Writing science: Literacy and discursive power.* London: Routledge.

Halliday, Michael Alexander Kirkwood / Christian Matthiessen 2004. *An Introduction to Functional Grammar.* London: Routledge.

Hasan, Ruqaiya 1995. The Conception of Context in Text. In Peter H. Fries / Michael Gregory (eds) *Discourse in Society: Systemic Functional Perspectives.* Norwood: Ablex, 183–284.

Haspelmath, Martin 2001. Word classes/parts of speech. In Neil J. Smelser / Paul B. Baltes (eds) *Encyclopedia of the Social and Behavioral Sciences.* Oxford: Pergamon, 16538–16545.

Haßler, Gerda 2015. Evidentiality and the expression of speaker's stance in Romance languages and German. *Discourse Studies* 17/2, 182–209.

Hiippala, Tuomo 2015. *The Structure of Multimodal Documents: An Empirical Approach.* London: Routledge.

Hoye, Leo 1997. *Adverbs and Modality in English.* New York: Routledge.

Holtz, Martin 2011. Lexico-grammatical properties of abstracts and research articles. A corpus-based study of scientific discourse from multiple disciplines. (Doctoral dissertation, Technischen Universität Darmstadt, Darmstadt, Germany).

Hu, Guangwei / Cao, Feng 2011. Hedging and boosting in abstracts of applied linguistics articles: A comparative study of English- and Chinese-medium journals. *Journal of Pragmatics* 43/11, 2795–2809.

Huddleston, Rodney / Pullum, Geoffrey K. 2002. *The Cambridge Grammar of the English Language.* Cambridge: Cambridge University Press.

Huddleston, Rodney / Pullum, Geoffrey 2017. *The Cambridge Grammar of the English Language.* Cambridge: Cambridge University Press.

Hunston, Susan / Thompson, Geoff (eds.) 2000. Evaluation in text: Authorial stance and the construction of discourse. Oxford: Oxford University Press.

Hyland, Ken 1998. *Hedging in Scientific Research Articles*. Amsterdam: John Benjamins.

Hyland, Ken 2004. Disciplinary interactions: Metadiscourse in L2 postgraduate writing. *Journal of Second Language Writing* 13/2, 133–151.

Hyland, Ken 2005. Stance and engagement: A model of interaction in academic discourse. *Discourse Studies* 7/2, 173–192.

Hyland, Ken / Jiang, Feng 2018. 'We believe that … ': Changes in an academic stance marker. *Australian Journal of Linguistics* 38/2, 139–161.

Hyland, Ken / Tse, Polly 2004. Metadiscourse in academic writing: A reappraisal. *Applied Linguistics* 25/2, 156–177.

Hyland, Ken / Tse, Polly 2005a. Evaluative that constructions: Signalling stance in research abstracts. *Functions of Language* 12/1, 39–63.

Hyland, Ken / Tse, Polly 2005b. Hooking the reader: A corpus study of evaluative that in abstracts. *English for Specific Purposes* 24/2, 123–139.

Hyon, Sunny 1996. Genre in the three traditions: Implications for ESL. *TESOL Quarterly* 30/4, 693–722.

Imao, Yasu 2020a. *CasualConc*. Retrieved from https://sites.google.com/site/casualconc/download (Accessed 13 July 2021)

Imao, Yasu 2020b. *CasualTextractor*. Retrieved from https://sites.google.com/site/casualconc/utility-programs/casualtextractor (Accessed 13 July 2021)

Jaime Pastor, María / Pérez-Guillot, Cristina 2015. A comparison analysis of modal auxiliary verbs in technical and general English. *Procedia - Social and Behavioral Sciences* 212, 292–297.

Jalilifar, Alireza / Hayati, Samira / Don, Alexanne 2018. Investigating metadiscourse markers in book reviews and blurbs: A study of interested and disinterested genres. *Studies About Languages* 2824/33, 90–107.

Johnstone, Barbara 2009. Stance, style, and the linguistic individual. In Alexandra Jaffe (ed.) *Stance: Sociolinguistic Perspectives*. Oxford: Oxford University Press, 29–71.

Kanoksilapatham, Budsaba 2005. Rhetorical structure of biochemistry research articles. *English for Specific Purposes* 24/3, 269–292.

Kawase, Tomoyuki 2015. Metadiscourse in the introductions of PhD theses and research articles. *Journal of English for Academic Purposes* 20, 114–124.

Kim, Chanhee / Crosthwaite, Peter 2019. Disciplinary differences in the use of evaluative that: Expression of stance via that-clauses in business and medicine. *Journal of English for Academic Purposes* 41, 1–14.

Kozáčiková, Zuzana 2021. Stance complement clauses controlled by verbs in academic research papers. *Topics in Linguistics* 22/1, 16–26.

Kranich, Svenja 2009. Epistemic modality in English popular scientific texts and their German translation. *Transkom* 2, 26–41.

Laghari, Tania / Akhter, Tahreem / Mastoi, Ruqia Bano 2021. A comparative analysis of applied linguistics and economic psychology research articles: Verb tense in conclusion section. *Pakistan Journal of Social Research* 3/4, 577–589.

Langacker, Ronald W. 2008. *Cognitive grammar. A basic introduction.* Oxford: Oxford University Press.

Langacker, Ronald W. 2010. Control and the mind/body duality: knowing vs. effecting. In Elżbieta Tabakowska / Michał Choiński / Łukasz Wiraszka (eds) *Cognitive Linguistics in Action: From Theory to Application and Back*, 165–207. (Applications of Cognitive Linguistics 14.) Berlin and New York: De Gruyter Mouton.

Lee, Seul-bi 2015. Discourse Functions of Tense-aspect Expression in Korean Research Articles. *Journal of Korean Language Education* 35, 161–198.

Leong, Ping Alvin 2014. The passive voice in scientific writing. The current norm in science journals. *Journal of Science Communication* 13/1 A03, 1–16.

Leong, Ping Alvin 2021. The passive voice in scholarly writing: A diachronic look at science and history. *Finnish Journal of Linguistics* 34, 77–102.

Li, Li Juan, / Ge, Guang Chun 2009. Genre analysis: Structural and linguistic evolution of the English-medium medical research article (1985–2004). *English for Specific Purposes* 28/2, 93–104.

Lin, Kathy Ling 2020. *Perspectives on the Introductory Phase of Empirical Research Articles: A Study of Rhetorical Structure and Citation Use.* Singapore: Springer.

Lin, Ling / Evans, Stephen 2012. Structural patterns in empirical research articles: A cross-disciplinary study. *English for Specific Purposes* 31/3, 150–160.

Liu, Yali / Buckingham, Louisa 2018. The schematic structure of discussion sections in applied linguistics and the distribution of metadiscourse markers. *Journal of English for Academic Purposes* 34, 97–109.

Loi, Chek Kim 2010. Research article introductions in Chinese and English: A comparative genre-based study. *Journal of English for Academic Purposes* 9/4, 267–279.

Lu, Xiaofei 2010. Automatic analysis of syntactic complexity in second language writing. *International Journal of Corpus Linguistics* 15/4, 474–496.

Lu, Xiaofei 2011. A corpus-based evaluation of syntactic complexity measures as indices of college-level ESL writers' language development. *TESOL Quarterly* 45/1, 36–62.

Lu, Xiaofei 2012. The relationship of lexical richness to the quality of ESL learners' oral narratives. *The Modern Language Journal* 96/2, 190–208.

Lu, Xiaofei 2014. *Computational methods for corpus annotation and analysis.* New York: Springer.

Lu, Xiaofei / Ai, Haiyang 2015. Syntactic complexity in college-level English writing: Differences among writers with diverse L1 backgrounds. *Journal of Second Language Writing* 29, 16–27.

Marín-Arrese, Juana Isabel 2009a. Commitment and subjectivity in the discourse of a judicial inquiry. In Raphael Salkie / Pierre Busuttil / Johan van der Auwera (eds) *Modality in English. theory and description.* Berlin: Mouton de Gruyter, 237–268.

Marín-Arrese, Juana Isabel 2009b. Effective vs. epistemic stance, and subjectivity/intersubjectivity in political discourse. A case study. In Anastasios Tsangalidis / Roberta Facchinetti (eds) *Studies on English modality. In honour of Frank R. Palmer.* Bern/New York: Peter Lang, 23–52.

Marín Arrese, Juana I. 2011. Effective vs. Epistemic stance and subjectivity in political discourse: Legitimising strategies and mystification of responsibility. In Christopher Hart (ed.) *Critical Discourse Studies in Context and Cognition.* Amsterdam: John Benjamins, 193–223.

Martin, James 1984. Language, register and genre. In Frances Christie (ed.) *Language Studies: Children's Writing: Reader.* Geelong: Deakin University Press, 21–29.

Martin, James 2000. Analysing genre: functional parameters. In Frances Christie / James R. Martin (eds) *Genre and Institutions. Social Processes in the Workplace and School.* London/New York: Continuum, 3–39.

Martin, James / Rose, David 2008. *Genre Relations. Mapping Culture.* London: Equinox.

Mauranen, Anna / Bondi, Marina 2003. Evaluative language use in academic discourse. *Journal of English for Academic Purposes* 2/4, 269–271.

Mongkholjuck, Chinanard 2008. A genre analysis of tourism attraction leaflet produced and distributed in Thailand in 2004. (Doctoral dissertation, Kasetsart University, Bangkok, Thailand).

Myers, Greg 1989. The pragmatics of politeness in scientific articles. *Applied Linguistics*, 10/1, 1–35.

Nadova, Zuzana 2015. Grammatical means of textual cohesion in appellate court decisions. *Revista de Lenguas para Fines Específicos* 21/2, 48–69.

Ngula, Richmond Sadick 2017. Epistemic modal verbs in research articles written by ghanaian and international scholars: A corpus-based study of three disciplines. *Brno Studies in English* 43/2, 5–27.

Nhat, Ton Nu My / Minh, Nguyen Thi Dieu 2020. A Study on Modality in English-Medium Research Articles. *VNU Journal of Foreign Studies* 36/6, 74–92.

Nuyts, Jan 2001. *Epistemic modality, language, and conceptualization.* Amsterdam: John Benjamins.

Nwogu, Kevin Ngozi 1997. The medical research paper: Structure and functions. *English for Specific Purposes* 16/2, 119–138.

Ochs, Elinor (ed.) 1989. *The pragmatics of affect.* Special issue of Text 9/1. Berlin: Mouton de Gruyter.

Okuyama, Yasuhiro 2020. Use of Tense and Aspect in Academic Writing in Engineering: Simple Past and Present Perfect. *Journal of Pan-Pacific Association of Applied Linguistics* 24/1, 1–15.

Paltridge, Brian 2001. *Genre and the language learning classroom.* Michigan: University of Michigan Press.

Palmer, Frank Robert 1986. *Mood and modality.* Cambridge: Cambridge University Press.

Palmer, Frank Robert 2001. *Mood and Modality.* Cambridge: Cambridge University Press.

Pic, Elsa / Furmaniak, Grégory 2012. A study of epistemic modality in academic and popularised discourse: The case of possibility adverbs perhaps, maybe and possibly. *Revista de Lenguas para Fines Específicos* 18, 13–44.

Piqué, Jordi / Andreu-Besó, J. Vicent 2000. A textual perspective of scientific articles: Patterns and moves. In Ruane, Mary / Baoill, Dónall Ó. (eds) *Integrating theory and practice in LSP and LAP.* Dublin: Applied Language Centre, University College Dublin & Irish Association of Applied Linguistics (IRAAL), 57–70.

Posteguillo, Santiago 1999. The schematic structure of computer science research articles. *English for Specific Purposes* 18/2, 139–160.

Puente Castelo, Luis M. / Monaco Leida, María 2013. Conditionals and their functions in women's scientific writing. *Procedia - Social and Behavioral Sciences* 95, 160–169.

Quirk, Randolph / Greenbaum, Sidney / Leech, Geoffrey / Svartvik, Jan 1972. *A Grammar of Contemporary English.* London: Longman.

Quirk, Randolph / Greenbaum, Sidney / Leech, Geoffrey / Svartik, Jan 1985. *A comprehensive grammar of the English language.* London: Longman.

Ramat, Paolo / Ricca, Davide 1998. Sentence adverbs in the languages of Europe. In In Johan Van der Auwera / Dónall Ó. Baoill / Donall PO Baoill (eds) *Adverbial constructions in the languages of Europe*, 187–275.

Reilly, Judy / Zamora, Anita / McGivern, Robert 2005. Acquiring perspective in English: The development of stance. *Journal of Pragmatics* 37/2, 185–208.

Ruelas Inzunza, Ernesto 2020. Reconsidering the Use of the Passive Voice in Scientific Writing. *The American Biology Teacher* 82/8, 563–565.

Ruffolo, Ida 2015. *The Perception of Nature in Travel Promotion Texts. A Corpus-based Discourse Analysis.* Bern: Peter Lang.

Ruiying, Yang / Allison, Desmond 2003. Research articles in applied linguistics: Moving from results to conclusions. *English for Specific Purposes* 22/4, 365–385.

Rundblad, Gabriella 2007. Impersonal, General, and Social. The Use of Metonymy Versus Passive Voice in Medical Discourse. *Written Communication* 24/3, 250–277.

Sabila, Nurul Akrima A. / Kurniawan, Eri 2020. Move analysis of tourism research article abstracts in national and international journal articles. *Proceedings of the 4th International Conference on Language, Literature, Culture, and Education (ICOLLITE 2020).* Paris, France: Atlantis Press, 514–520.

Sepehri, Mehrdad / Hajijalili, Mehrnnoosh / Namaziandost, Ehsan 2019. Hedges and boosters in medical and engineering research articles: A comparative corpus-based study. *Global Journal of Foreign Language Teaching* 9/4, 215–225.

Stoller, Fredricka L. / Robinson, Marin 2013. Chemistry journal articles: An interdisciplinary approach to move analysis with pedagogical aims. *English for Specific Purposes* 32/1, 45–57.

Stotesbury, Hilkka 2003. Evaluation in research article abstracts in the narrative and hard sciences. *Journal of English for Academic Purposes* 2/4, 327–341.

Swales, John M. 1990. *Genre analysis.* Cambridge: Cambridge University Press.

Swales, John M. 2004. *Research genres. Explorations and applications.* Cambridge: Cambridge University Press.

Sweetser, Eve 1990. *From Etymology to Pragmatics. Metaphorical and Cultural Aspects of Semantic Structure.* Cambridge: Cambridge University Press.

Taillon, Justin 2014. Understanding tourism as an academic community, study or discipline. *Journal of Tourism & Hospitality* 3/3, 1–5.

Tardy, Christine M. / Swales, John M. 2014. Genre analysis. In Klaus P. Schneider / Anne Barron (eds) *Pragmatics of discourse*, 165–188.

Tessuto, Girolamo 2015. Generic structure and rhetorical moves in English-language empirical law research articles: Sites of interdisciplinary and interdiscursive cross-over. *English for Specific Purposes* 37, 13–26.

Van Gelderen, Elly 2010. *An Introduction to the Grammar of English.* Amsterdam: John Benjamins.

Van der Auwera, Johan and Plungian, Vladimir A. 1998. On modality's semantic map. *Linguistic Typology* 2, 79–124.

Vázquez Orta, Ignacio 2010. A contrastive analysis of the use of modal verbs in the expression of epistemic stance in business management research articles in English and Spanish. *Iberica* 19, 77–96.

Wang, Shih-ping / Tu, Pin-ning 2014. Tense use and move analysis in journal article abstracts. *Taiwan Journal of TESOL* 11/1, 3–29.

Warchał, Krystyna 2010. Moulding interpersonal relations through conditional clauses: Consensus-building strategies in written academic discourse. *Journal of English for Academic Purposes* 9/2, 140–150.

Weissberg, Robert / Buker, Suzanne 1990. *Writing up research: Experimental research report writing for students of English*. Englewood Cliffs, NJ: Prentice Hall Regents.

Willet, Thomas 1988. A cross-linguistic survey of the grammaticalization of evidentiality. *Studies in Language* 12/1, 51–97.

Yui Ling Ip, Janice 2008. Analyzing tourism discourse: A case study of a Hong Kong travel brochure. *LOCOM Papers* 1, 1–19.

Appendix: Corpus references

Aquino, Richard S. / Schänzel, Heike A. / Hyde, Kenneth F. 2018. Unearthing the geotourism experience: Geotourist perspectives at Mount Pinatubo, Philippines. *Tourist Studies* 18/1, 41–62. http://doi:10.1177/1468797617717465.

Atadil, Hilmi A. / Sirakaya-Turk, Ercan / Altintas, Volkan 2017. An analysis of destination image for emerging markets of Turkey. *Journal of Vacation Marketing* 23/1, 37–54. http://doi:10.1177/1356766715616858.

Aukland, Knut 2016. Retailing religion: Guided tours and guide narratives in Hindu pilgrimage. *Tourist Studies* 16/3, 237–257. http://doi:10.1177/1468797615618038.

Barry, Kaya 2017. Diagramming: A creative methodology for tourist studies. *Tourist Studies* 17/3, 328–346. http://doi:10.1177/146879761 6680852.

Beritelli, Pietro / Reinhold, Stephan 2018. Chance meetings, the destination paradox, and the social origins of travel: Predicting traveler's whereabouts? *Tourist Studies* 18/4, 417–441. http://doi:10.1177/1468797617748292.

Bezzola, Toya / Lugosi, Peter 2018. Negotiating place through food and drink: Experiencing home and away. *Tourist Studies* 18/4, 486–506. http://doi:10.1177/1468797618791125.

Bolderman, Leonieke / Reijnders, Stijn 2017. Have you found what you're looking for? Analysing tourist experiences of Wagner's Bayreuth, ABBA's Stockholm and U2's Dublin. *Tourist Studies* 17/2, 164–181. http://doi:10.1177/1468797616665757.

Bone, Jane / Bone, Kate 2018. Voluntourism as cartography of self: A Deleuzian analysis of a postgraduate visit to India. *Tourist Studies* 18/2, 177–193. http://doi:10.1177/1468797617723468.

Bratt, Jonathan 2018. Return to the east: Tourism promotion as legitimation in Qiandongnan, China. *Tourist Studies* 18/1, 21–40. http://doi:10.1177/1468797617711575.

Bruttomesso, Elisa 2018. Making sense of the square: Facing the touristification of public space through playful protest in Barcelona. *Tourist Studies* 18/4, 467–485. http://doi:10.1177/1468797618775219.

Cashman, David 2017. 'The most atypical experience of my life': The experience of popular music festivals on cruise ships. *Tourist Studies* 17/3, 245–262. http://doi:10.1177/1468797616665767.

Chapman, Anya / Light, Duncan 2017. Working with the carnivalesque at the seaside: Transgression and misbehaviour in a tourism workplace. *Tourist Studies* 17/2, 182–199. http://doi:10.1177/146879761 6665768.

Chen, De Jung 2018. Couchsurfing: Performing the travel style through hospitality exchange. *Tourist Studies* 18/1, 105–122. http://doi:10.1177/1468797617710597.

Cooke, Lisa 2017. Carving "turns" and unsettling the ground under our feet (and skis): A reading of Sun Peaks Resort as a settler colonial moral terrain. *Tourist Studies* 17/1, 36–53. http://doi:10.1177/14687 97616685643.

Correia, Antonia / Kozak, Metin / Kim, Seongseop (Sam) 2018. Luxury shopping orientations of mainland Chinese tourists in Hong Kong: Their shopping destination. *Tourism Economics* 24/1, 92–108. http://doi:10.1177/1354816617725453.

Country, Bawaka / Wright, Sarah / Lloyd, Kate / Suchet-Pearson, Sandie / Burarrwanga, Laklak / Ganambarr, Ritjilili / Ganambarr, Merrkiyawuy / Ganambarr, Banbapuy / Maymuru, Djawundil / Tofa, Matalena 2017. Meaningful tourist transformations with Country at Bawaka, North East Arnhem Land, northern Australia. *Tourist Studies* 17/4, 443–467. http://doi:10.1177/1468797616682134.

Dumbrăveanu, Daniela / Light, Duncan / Young, Craig / Chapman, Anya 2016. Exploring women's employment in tourism under state socialism: Experiences of tourism work in socialist Romania. *Tourist Studies* 16/2, 151–169. http://doi:10.1177/1468797615594747.

Fillis, Ian / Lehman, Kim / Miles, Morgan P. 2015. The museum of old and new art: Leveraging entrepreneurial marketing to create a unique arts and vacation venture. *Journal of Vacation Marketing* 23/1, 85–96. http://doi:10.1177/1356766716634153.

Fitzgerald, Jon / Reis, Arianne C. 2016. Island intersections: Music and tourism on Fernando de Noronha, Brazil. *Tourist Studies* 16/2, 170–191. http://doi:10.1177/1468797615594738.

Fredman, Peter / Wikström, Daniel 2018. Income elasticity of demand for tourism at Fulufjället National Park. *Tourism Economics* 24/1, 51–63. http://doi:10.1177/1354816617724012.

Garcia, Luis Manuel 2016. Techno-tourism and post-industrial neo-romanticism in Berlin's electronic dance music scenes. *Tourist Studies* 16/3, 276–295. http://doi:10.1177/1468797615618037.

Garner, Ross 2017. Insecure positions, heteronomous autonomy and tourism-cultural capital: A Bourdieusian reading of tour guides on BBC Worldwide's Doctor Who Experience Walking Tour. *Tourist Studies* 17/4, 426–442. http://doi:10.1177/1468797616680851.

Garrigos-Simon, Fernando J. / Llorente, Roberto / Morant, Maria / Narangajavana, Yeamduan 2016. Pervasive information gathering and data mining for efficient business administration. *Journal of Vacation Marketing* 22/4, 295–306. http://doi:10.1177/135676671 5617219.

Gillen, Jamie 2016. Urbanizing existential authenticity: Motorbike tourism in Ho Chi Minh City, Vietnam. *Tourist Studies* 16/3, 258–275. http://doi:10.1177/1468797615618035.

Gothie, Sarah Conrad 2016. Playing "Anne": Red braids, Green Gables, and literary tourists on Prince Edward Island. *Tourist Studies* 16/4, 405–421. http://doi:10.1177/1468797615618092.

Gyimóthy, Szilvia 2018. Transformations in destination texture: Curry and Bollywood romance in the Swiss Alps. *Tourist Studies* 18/3, 292–314. http://doi:10.1177/1468797618771692.

Hales, Rob / Caton, Kellee 2017. Proximity ethics, climate change and the flyer's dilemma: Ethical negotiations of the hypermobile traveller. *Tourist Studies* 17/1, 94–113. http://doi:10.1177/146879761 6685650.

Hermann, Inge / Peters, Karin / Van Trijp, Emy 2017. Enrich yourself by helping others: A web content analysis of providers of gap year packages and activities in the Netherlands. *Tourist Studies* 17/1, 75–93. http://doi:10.1177/1468797616685649.

Hocking, Bree T. 2016. Gazed and subdued? Spectacle, spatial order and identity in the contested city. *Tourist Studies* 16/4, 367–385. http://doi:10.1177/1468797615618124.

Huang, Leo / Chang, Michael 2018. Why do travel agencies choose to undergo IPOs in Taiwan? *Tourism Economics* 24/1, 79–91. http://doi:10.1177/1354816617725452.

Jarratt, David / Sharpley, Richard 2017. Tourists at the seaside: Exploring the spiritual dimension. *Tourist Studies* 17/4, 349–368. http://doi:10.1177/1468797616687560.

Jethro, Duane 2016. 'Freedom Park, A Heritage Destination': Tour-guiding and visitor experience at a post-apartheid heritage site. *Tourist Studies* 16/4, 446–461. http://doi:10.1177/1468797615618099.

Jiménez-Esquinas, Guadalupe / Sánchez-Carretero, Cristina 2018. Who owns the name of a place? On place branding and logics in two villages in Galicia, Spain. *Tourist Studies* 18/1, 3–20. http://doi:10.1177/1468797617694728.

Johinke, Rebecca 2018. Take a walk on the wild side: Punk music walking tours in New York City. *Tourist Studies* 18/3, 315–331. http://doi:10.1177/1468797618771694.

Jönsson, Erik 2016. The nature of an upscale nature: Bro Hof Slott Golf Club and the political ecology of high-end golf. *Tourist Studies* 16/3, 315–336. http://doi:10.1177/1468797615618306.

Kawashima, Tinka Delakorda 2015. Travel agencies and priests as spiritual leaders: The merits of collaboration. *Tourist Studies* 16/1, 40–56. http://doi:10.1177/1468797615588430.

Liston-Heyes, Catherine / Daley, Carol 2017. Voluntourism, sensemaking and the leisure-volunteer duality. *Tourist Studies* 17/3, 283–305. http://doi:10.1177/1468797616665769.

Lund, Katrín Anna / Kjartansdóttir, Katla / Loftsdóttir, Kristín 2018. "Puffin love": Performing and creating Arctic landscapes in Iceland through souvenirs. *Tourist Studies* 18/2, 142–158. http://doi:10.1177/1468797617722353.

Lund, Katrín Anna / Loftsdóttir, Kristín / Leonard, Michael 2017. More than a stopover: Analysing the postcolonial image of Iceland as a gateway destination. *Tourist Studies* 17/2, 144–163. http://doi:10.1177/1468797616659951.

Lundberg, Christine / Ziakas, Vassilios / Morgan, Nigel 2018. Conceptualising on-screen tourism destination development. *Tourist Studies* 18/1, 83–104. http://doi:10.1177/1468797617708511.

Martín Rodríguez, Andrea / O'Connell, John F. 2018. Can low-cost long-haul carriers replace Charter airlines in the long-haul market? A European perspective. *Tourism Economics* 24/1, 64–78. http://doi:10.1177/1354816617724017.

Mendieta-Peñalver, Luis Felipe / Perles-Ribes, José F. / Ramón-Rodríguez, Ana B. / Such-Devesa, María J. 2018. Is hotel efficiency necessary for tourism destination competitiveness? An integrated approach. *Tourism Economics* 24/1, 3–26. http://doi:10.5367/te.2016.0555.

Nelson, Kim / Matthews, Amie Louise 2018. Foreign presents or foreign presence? Resident perceptions of Australian and Chinese tourists in Niseko, Japan. *Tourist Studies* 18/2, 213–231. http://doi:10.1177/1468797617717466.

Nilsson, Mats / Tesfahuney, Mekonnen 2018. The post-secular tourist: Re-thinking pilgrimage tourism. *Tourist Studies* 18/2, 159–176. http://doi:10.1177/1468797617723467.

Park, Hyejin, Soobin Seo / Kandampully, Jay 2016. Why post on social networking sites (SNS)? Examining motives for visiting and sharing pilgrimage experiences on SNS. *Journal of Vacation Marketing* 22/4, 307–319. http://doi:10.1177/1356766715615912.

Payne, James E. / Gil-Alana, Luis A. 2018. Data measurement and the change in persistence of tourist arrivals to the United States in the aftermath of the September 11th terrorist attacks. *Tourism Economics* 24/1, 41–50. http://doi:10.1177/1354816617719161.

Pham, Tien Duc / Nghiem, Son / Dwyer, Larry 2018. The economic impacts of a changing visa fee for Chinese tourists to Australia. *Tourism Economics* 24/1, 109–126. http://doi:10.1177/135481661 7726204.

Prince, Solène 2018. Dwelling in the tourist landscape: Embodiment and everyday life among the craft-artists of Bornholm. *Tourist Studies* 18/1, 63–82. http://doi:10.1177/1468797617710598.

Reichenberger, Ina 2017. Why the host community just isn't enough: Processes and impacts of backpacker social interactions. *Tourist Studies* 17/3, 263–282. http://doi:10.1177/1468797616665770.

Reitsamer, Bernd Frederik / Brunner-Sperdin, Alexandra 2017. Tourist destination perception and well-being: What makes a destination attractive? *Journal of Vacation Marketing* 23/1, 55–72. http://doi:10.1177/1356766715615914.

Rickly, Jillian M. 2017. "I'm a Red River local": Rock climbing mobilities and community hospitalities. *Tourist Studies* 17/1, 54–74. http://doi:10.1177/1468797616685648.

Ryan, Louise 2016. Re-branding Tasmania: MONA and the altering of local reputation and identity. *Tourist Studies* 16/4, 422–445. http://doi:10.1177/1468797615618097.

Salvatierra, Javier / Walters, Gabrielle 2015. The impact of human-induced environmental destruction on destination image perception and travel behaviour: The case of Australia's Great Barrier

Reef. *Journal of Vacation Marketing* 23/1, 73–84. http://doi:10.1177/1356766715626966.

Sampaio, Sofia 2017. Tourism, gender and consumer culture in late- and post-authoritarian Portugal. *Tourist Studies* 17/2, 200–217. http://doi:10.1177/1468797616665771.

Schäfer, Stefanie 2016. From Geisha girls to the Atomic Bomb Dome: Dark tourism and the formation of Hiroshima memory. *Tourist Studies* 16/4, 351–366. http://doi:10.1177/1468797615618122.

Schänzel, Heike A. / Lynch, Paul A. 2016. Family perspectives on social hospitality dimensions while on holiday. *Tourist Studies* 16/2, 133–150. http://doi:10.1177/1468797615594742.

Shiu, Jerry Yuwen 2018. Individual rationality and differences in Taiwanese spa hotel choice. *Tourism Economics* 24/1, 27–40. http://doi:10.1177/1354816617718972.

Skinner, Jonathan 2015. Walking the Falls: Dark tourism and the significance of movement on the political tour of West Belfast. *Tourist Studies* 16/1, 23–39. http://doi:10.1177/1468797615588427.

Straub, Leslie Ellen 2015. Negotiation and experience: Space and place in religious pilgrimage. *Tourist Studies* 16/1, 88–104. http://doi:10.1177/1468797616635378.

Tasci, Asli D.A. / Jae Ko, Yong 2017. Travel needs revisited. *Journal of Vacation Marketing* 23/1, 20–36. http://doi:10.1177/1356766715617499.

Tchoukarine, Igor 2016. A Place of Your Own on Tito's Adriatic: Club Med and Czechoslovak Trade Union Holiday Resorts in the 1960s. *Tourist Studies* 16/4, 386–404. http://doi:10.1177/1468797615618125.

Tegtmeyer, Lina L. 2016. Tourism aesthetics in ruinscapes: Bargaining cultural and monetary values of Detroit's negative image. *Tourist Studies* 16/4, 462–477. http://doi:10.1177/1468797615618100.

Torabian, Pooneh / Mair, Heather 2017. (Re)constructing the Canadian border: Anti-mobilities and tourism. *Tourist Studies* 17/1, 17–35. http://doi:10.1177/1468797616685645.

Tzanelli, Rodanthi / Korstanje, Maximiliano E. 2016. Tourism in the European economic crisis: Mediatised worldmaking and new tourist imaginaries in Greece. *Tourist Studies* 16/3, 296–314. http://doi:10.1177/1468797616648542.

Valenta, Marko / Strabac, Zan 2016. The dramaturgical nexus of ethno-religious, tourist and transnational frames of pilgrimages in post-conflict societies: The Bosnian and Herzegovinian experience. *Tourist Studies* 16/1, 57–73. http://doi:10.1177/1468797616635371.

van Nuenen, T. (2016). Here I am: Authenticity and self-branding on travel blogs. *Tourist Studies* 16/2, 192–212. http://doi.org/10.1177/1468797615594748.

Vanolo, Alberto / Cattan, Nadine 2017. Selling cruises: Gender and mobility in promotional brochures. *Tourist Studies* 17/4, 406–425. http://doi:10.1177/1468797616682615.

Vorobjovas-Pinta, Oskaras / Robards, Brady 2017. The shared   An insider ethnographic account of a gay resort. *Tourist Studies* 17/4, 369–387. http://doi:10.1177/1468797616687561.

Walter, Pierre G. 2016. Travelers' experiences of authenticity in "hill tribe" tourism in Northern Thailand. *Tourist Studies* 16/2, 213–230. http://doi:10.1177/1468797615594744.

Wang, Li / Alasuutari, Pertti 2017. Co-construction of the tourist experience in social networking sites: Two forms of authenticity intertwined. *Tourist Studies* 17/4, 388–405. http://doi:10.1177/1468797616687559.

Weatherby, Theodora G. / Vidon Elizabeth S. 2018. Delegitimizing wilderness as the man cave: The role of social media in female wilderness empowerment. *Tourist Studies* 18/3, 332–352. http://doi:10.1177/1468797618771691.

White, Leanne 2018. Qantas still calls Australia home: The spirit of Australia and the flying kangaroo. *Tourist Studies* 18/3, 261–274. http://doi:10.1177/1468797618785617.

Zerva, Konstantina 2018. 'Chance Tourism': Lucky enough to have seen what you will never see. *Tourist Studies* 18/2, 232–254. http://doi:10.1177/1468797617723471.

Zhang, Xiaoyu / Ryan Chris 2018. Investigating tourists' and local residents' perceptions of a Chinese film site. *Tourist Studies* 18/3, 275–291. http://doi:10.1177/1468797618771693.

Linguistic Insights

Studies in Language and Communication

This series aims to promote specialist language studies in the fields of linguistic theory and applied linguistics, by publishing volumes that focus on specific aspects of language use in one or several languages and provide valuable insights into language and communication research. A cross-disciplinary approach is favoured and most European languages are accepted.

The series includes two types of books:

– Monographs – featuring in-depth studies on special aspects of language theory, language analysis or language teaching.
– Collected papers – assembling papers from workshops, conferences or symposia.

Each volume of the series is subjected to a double peer-reviewing process.

Vol. 1   Maurizio Gotti & Marina Dossena (eds)
Modality in Specialized Texts. Selected Papers of the 1st CERLIS Conference.
421 pages. 2001. ISBN 3-906767-10-8 · US-ISBN 0-8204-5340-4

Vol. 2   Giuseppina Cortese & Philip Riley (eds)
Domain-specific English. Textual Practices across Communities
and Classrooms.
420 pages. 2002. ISBN 3-906768-98-8 · US-ISBN 0-8204-5884-8

Vol. 3   Maurizio Gotti, Dorothee Heller & Marina Dossena (eds)
Conflict and Negotiation in Specialized Texts. Selected Papers
of the 2nd CERLIS Conference.
470 pages. 2002. ISBN 3-906769-12-7 · US-ISBN 0-8204-5887-2

Vol. 4   Maurizio Gotti, Marina Dossena, Richard Dury, Roberta Facchinetti & Maria Lima
Variation in Central Modals. A Repertoire of Forms and Types of Usage
in Middle English and Early Modern English.
364 pages. 2002. ISBN 3-906769-84-4 · US-ISBN 0-8204-5898-8

Editorial address:

Prof. Maurizio Gotti   Università di Bergamo, Dipartimento di Lingue, Letterature e Culture
Straniere Piazza Rosate 2, 24129 Bergamo, Italy
Fax: +39 035 2052789, E-Mail: m.gotti@unibg.it